Lecture Notes in Artificial Intelligence 11325

Subseries of Lecture Notes in Computer Science

More information about this series at http://www.springer.com/series/1244

Wei Lee Woon · Zeyar Aung
Alejandro Catalina Feliú · Stuart Madnick (Eds.)

Data Analytics for Renewable Energy Integration

Technologies, Systems and Society

6th ECML PKDD Workshop, DARE 2018
Dublin, Ireland, September 10, 2018
Revised Selected Papers

 Springer

Editors
Wei Lee Woon
Masdar Institute (Khalifa University)
Abu Dhabi, United Arab Emirates

Zeyar Aung
Masdar Institute (Khalifa University)
Abu Dhabi, United Arab Emirates

Alejandro Catalina Feliú
Autonomous University of Madrid
Madrid, Spain

Stuart Madnick
Massachusetts Institute of Technology
Cambridge, MA, USA

ISSN 0302-9743 ISSN 1611-3349 (electronic)
Lecture Notes in Artificial Intelligence
ISBN 978-3-030-04302-5 ISBN 978-3-030-04303-2 (eBook)
https://doi.org/10.1007/978-3-030-04303-2

Library of Congress Control Number: 2018961001

LNCS Sublibrary: SL7 – Artificial Intelligence

This Springer imprint is published by the registered company Springer Nature Switzerland AG
The registered company address is: Gewerbestrasse 11, 6330 Cham, Switzerland

Preface

This volume presents a collection of papers focused on the use of data analytics and machine learning techniques to facilitate the integration of renewable energy resources into existing infrastructure and socioeconomic systems. This collection includes papers presented at DARE 2018: the 6th International Workshop on Data Analytics for Renewable Energy Integration, which was hosted by ECML PKDD 2018.

Climate change, energy security, and sustainability have stimulated extensive research on the development of clean and renewable energy sources. However, of equal importance is the integration of these sources into existing infrastructure and socioeconomic systems. While increasing the generating capacities of renewable energy sources is still important, issues such as efficient and cost-effective storage and distribution, demand response, planning, and policy making must be resolved in parallel. These challenges are inherently multidisciplinary and depend heavily on robust and scalable computing techniques and the ability to handle large, complex data sets. The fields of data analytics, pattern recognition, and machine learning have a lot to offer in this context. Relevant topics include fault and event detection, forecasting of energy time series, cyber security, and demand response.

An additional element that was highlighted this year was the integration of renewable energy in society. It is clear that the spread of renewable energy will require the participation and support of end users. In this context, related topics include demand response, residential PV installations and even social media analytics in the context of building and measuring awareness of and attitudes toward renewable energy. The focus of this workshop is to study and present the use of various data analytics techniques in the different areas of renewable energy integration.

This year's event attracted numerous researchers working in the various related domains, both to present and discuss their findings and to share their respective experiences and concerns. We are very grateful to the organizers of ECML PKDD 2018 for hosting DARE 2018, the Program Committee members for their time and assistance, and the Masdar Institute, MIT, and the Universidad Autónoma de Madrid for their support of this timely and important workshop. Last but not least, we sincerely thank the authors for their valuable contributions to this volume.

October 2018

Wei Lee Woon
Zeyar Aung
Alejandro Catalina Feliú
Stuart Madnick

Organization

Program Chairs

Wei Lee Woon   Masdar Institute (Khalifa University of Science
                                  and Technology), UAE
Zeyar Aung   Masdar Institute of Science and Technology, UAE
Alejandro Catalina Feliú   Universidad Autónoma de Madrid, Spain
Stuart Madnick   Massachusetts Institute of Technology, USA

Program Committee

Elie Azar   Khalifa University of Science and Technology, UAE
João Catalão   University of Beira Interior, Portugal
Mustafa Amir Faisal   University of Texas at Dallas, USA
Anne Kayem   Hasso-Plattner Institute, Germany
Xiaoli Li   Institute for Infocomm Research, Singapore
Daisuke Mashima   Advanced Digital Sciences Center, Singapore
Bijay Neupane   Aalborg University, Denmark
Taha Ouarda   National Institute of Scientific Research, Canada
Jimmy Peng   National University of Singapore, Singapore
Filipe Saraiva   Federal University of Para, Brazil
Moamar Sayed-Mouchaweh   Institute Mines-Telecom Lille Douai, France
Weidong Xiao   University of Sydney, Australia
Haiwang Zhong   Tsinghua University, China

Contents

Mathematical Optimization of Design Parameters of Photovoltaic Module

Dávid Kubík$^{(\boxtimes)}$ and Jaroslav Loebl

Faculty of Informatics and Information Technologies, Slovak University
of Technology in Bratislava, Ilkovičova 2, 842 16 Bratislava 4, Slovakia
kubik.david5@gmail.com, jaroslavloebl@gmail.com
https://www.fiit.stuba.sk

Abstract. We propose a method for finding the most appropriate photovoltaic (hereafter PV) module and the return on investment for specific household needs, leveraging mathematical optimization. Based on
electricity consumption and location of the household, the algorithm
finds PV module design parameters using Covariance Matrix Adaptation
Evolution Strategy (hereafter CMA-ES). According to these computed
design parameters, the algorithm finds the most similar PV module from
the dataset of PV modules using Euclidean distance. Subsequently, savings and costs are calculated for the recommended PV system. When
calculating the return on investment, the algorithm takes into account
the price of electricity, initial investment and battery replacements. The
objective function is defined to maximize net savings during the lifetime of the PV system, which is considered to be 20 years. Heuristics
of the battery takes into account the aging of batteries and the loss of
energy due to storage. Battery life is set at 10 years, as we are considering using high quality batteries. In order to extend battery life, the
algorithm counts with the charge level restrictions.

Keywords: Mathematical optimization · Design parameters
Photovoltaic module

1 Introduction

The economy of the PV system, for ordinary users, is highly dependent on their
consumer habits and lifestyle, as it affects the electricity consumption curve
throughout the day. The ideal case would be if consumers spent their energy
just in time when the PV module generates the most electricity. However, this
ideal scenario is out of the question in practice because people are usually at work
at a given time unless it is a weekend or a day off. This problem can be solved
to a considerable extent by installing batteries that are charged at the peak of
electricity generation, and then this energy can be used to cover increased consumption at a time when the PV module cannot fully cover household demand.

© Springer Nature Switzerland AG 2018
W. L. Woon et al. (Eds.): DARE 2018, LNAI 11325, pp. 1–12, 2018.
https://doi.org/10.1007/978-3-030-04303-2_1

2   Related Work and Motivation

In recent years, interest in renewable energy has grown and it is natural that a number of projects have been created to help calculate and decide whether the investment in the PV system will return. The most well-known projects are Project Sunroof developed by Google [10] and System Advisor Model [6] developed by the National Renewable Energy Laboratory.

The main aim of the Project Sunroof [10] is to provide the user with information about the solar potential of his roof. The biggest drawback of the Project Sunroof is that it only takes into account some areas of the United States and no other country. Also, a user cannot choose a more detailed consumption curve or at least some model household, he can only enter an electricity bill. Even the selections of the PV module or the size of the battery capacity are not available.

System Advisor Model [6] is software that provides a performance and finance model for renewable energy. In the field of renewable energy, this software covers photovoltaics, wind turbines and even biomass. Unlike the Project Sunroof, they allow the user to choose the individual elements of the system from their component database. They also allow the user to select the weather for a particular area from the list, download weather data from the internet or create their own weather data. On the other hand, the software is very extensive and therefore the orientation in it is not easy at all. With the huge amount of functionality the software provides, the user can quickly lose himself.

Neither of these applications will recommend a specific PV module to the client. We have implemented a mechanism to recommend a particular PV module from the PV modules database and we consider it a key feature of our application. We decided to draw inspiration from the aforementioned projects and incorporate their shortcomings. Especially, we wanted our user interface to be intuitive and easy to use. Our application also allows the user to adjust the electricity consumption curve and input the number of photovoltaic modules and batteries. The application is free and is not limited to any area.

3   Model for Finding an Optimal PV Module

Our goal is to maximize the net savings by means of optimization function defined by Eq. 1. Based on annual data on electricity consumption, electricity prices and location, the algorithm decides whether the investment will pay off. The objective function to maximize net savings is defined as follows:

$$\text{maximize } S_n(X) = 20 * S_a(X) - C_t \qquad (1)$$

where S_n represents net savings, $S_a(X)$ represents average annual savings saved by the PV system with or without battery, X represents PV module design parameters to be optimized (and azimuth if chosen), average annual savings are multiplied by 20 because we assume the system lifetime to be 20 years without the need for a complete component replacement, C_t represents total installation costs of the PV system and is defined by the following formula:

$$C_t = 2 * (P_b + I_b) + P_c + I_c + P_i + P_m + I_{imw} \qquad (2)$$

where P_b represents the price for battery purchase, I_b represents the cost of installing batteries, P_c represents the price of charge controller, I_c represents the cost of installing the charge controller, P_i represents the price of solar inverter, P_m represents the amount of investment in PV modules, I_{imw} represents the cost of installing solar inverter, PV modules and wiring costs. The amount of investment in batteries and the cost of installing batteries are multiplied by 2, because we expect the battery lifetime to be 10 years, but the lifetime of the PV system to be 20 years.

Due to the fact that the battery voltage affects the type of charge controller, we have decided to choose the appropriate battery type and the charge controller that work together. We have given the user the ability to enter the number of batteries. We chose the battery [9] with a price of €425 after the price was converted from the dollars. The appropriate charge controller [8] has a price €257.

The price of the inverter depends on the power of the PV system. After consulting the prices with the company ESolar [3], we chose the following prices shown in Table 1.

Table 1. Solar inverter prices based on PV system power.

Power (W)	Price (€)
1–1500	520
1501–2000	590
2001–2500	650
2501–4000	1090
4001–10000	2560

We have also consulted the installation costs with the company ESolar [3] and came up with the following prices shown in Table 2:

Table 2. Prices of metal construction for mounting PV modules depending on the location of the PV modules.

Placement location (W)	Price per PV module (€)
Pitched roof	53
Facade	86

The price of installation of the PV modules and solar inverter, including the wiring, is calculated as €0.29 per 1 Wp of PV system performance. Estimated cost of battery installation is €15 per battery and estimated cost of charge controller installation is €10. Since databases of PV modules do not include their average prices, the prices of PV modules are estimated as follows:

$$P_m = n * M_{pp} * P_{wp} \tag{3}$$

where P_m represents the amount of investment in PV modules, n represents the number of PV modules, M_{pp} represents PV module peak performance, P_{wp} represents price for Wp (watt peak). Due to the huge variety of PV modules and the prices, the price for Wp can be changed by the user. The default value is arbitrarily set to €0.65.

The choice of a particular PV module from the dataset is calculated as the least different PV module from the PV module design parameters obtained by the optimization. Finding the least different PV module is done by computing the Euclidean distance between design parameters of PV modules from dataset and design parameters obtained by optimization. As the least different PV module, we consider the one whose Euclidean distance is the smallest.

3.1   Battery Model

If the initial costs of purchasing batteries are not too high, PV battery systems become more economical choice [5,11]. To minimize battery replacement costs, it is necessary to keep the battery life as long as possible. We consider using a 12 V (volt) lithium-based battery system with 10 years lifetime warranty because they are industry standard [1]. In the paper [11], the batteries are modelled with a simplified approach, considering a mean watt-hour efficiency of 95%. For that reason, we have introduced a constant L that expresses the loss of energy caused by battery storage. By doing so, the battery will drain faster than its output to the appliance network (1.0526 Wh of battery capacity is consumed for each Wh covered consumption, it is in fact a reversed value of mean watt-hour efficiency and we got it from the ratio of 100%/95%). They also restricted the state of charge of the battery to a range between 20% and 80% of the nominal battery capacity to extend the battery life to the maximum. Therefore, the battery capacity with which the algorithm will work will be 0.6 * total battery capacity.

The battery will be charged exclusively from the PV module and will only be charged when there is excess of amount of generated electricity. The battery will be discharged at a time when the PV module will not cover all consumption. The battery will be discharged until there will be a need to cover consumption or until the state of charge reaches zero. With the state of charge being equal to 0, the battery still holds the amount of energy equal to 20% of its total capacity.

Basically, the ageing of the battery leads to a decrease in battery capacity. Therefore, it is assumed that 90% of the usable battery capacity is indeed utilised within the lifetime on average.

Battery capacity is defined by the following formula:

$$C_{Wh} = 0.9 * C_{Ah} * V \tag{4}$$

where C_{Wh} represents battery capacity in Wh (watt hour), C_{Ah} represents battery capacity in Ah (ampere hour) and V represents voltage of a battery in volts. We multiply this equation by 0.9 because of the ageing of the battery as discussed above.

Battery capacity with a restriction of charge and discharge to a defined level is calculated as follows:

$$C_{rWh} = C_{Wh} * (1 - (R_L + (1 - R_H))) \tag{5}$$

where C_{rWh} represents restricted battery capacity in Wh, R_L represents the battery charge level below which the battery must not be discharged and R_H represents the battery charge level above which the battery must not be charged.

Charging the battery is defined by the following formula:

$$SOC = SOC + ((W_c * h_c)/C_{rWh}) \tag{6}$$

where SOC represents state of charge of the battery, W_c represents charging the battery in watts and it is multiplied by h_c which represents the number of hours the battery is charged. It is divided by C_{rWh} so that when the SOC is equal to 0, the battery still holds the amount of energy equal to 20% of its total capacity. Similarly, when SOC is equal to 1, the battery holds the amount of energy equal to 80% of its total capacity.

Discharging the battery is defined by the following formula:

$$SOC = SOC - ((W_d * h_d * L)/C_{rWh}) \tag{7}$$

where W_d represents discharging the battery in watts, h_d represents the number of hours the battery is discharged and L is our introduced constant that expresses the loss of energy caused by battery storage and is equal to 1.0526 as mentioned above.

3.2 Optimization Method

We used Covariance Matrix Adaptation Evolution Strategy [4] (hereafter CMA-ES) for optimization. The evolution strategy is a stochastic, derivative-free search algorithm and therefore is suitable for maximizing/minimizing the non-linear objective function, which is our case (maximizing net savings). Because it is an evolutionary algorithm, it is based on the principle of biological evolution (recombination, mutation and selection of individuals). The covariance matrix is used for representing pairwise dependencies between the variables in the mutation distribution. The method for updating these variables is called Covariance Matrix Adaptation.

All design parameters were scaled to a range <0, 1>. We set the initialization vector to value 0.5 for each scaled design parameter. In each iteration of the mathematical optimization, the design parameters of the ideal PV module are modified and sent to objective function which is shown in the activity diagram 3.2 (Fig. 1).

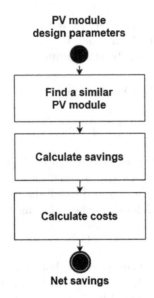

Fig. 1. Activity diagram showing the flow of the objective function.

The goal of mathematical optimization is to find a PV system that will have the highest net savings. Extensive testing has shown, that the ideal setting of the generation number is 10, in terms of the net savings/the duration of the calculation ratio.

4 Results

The input data varied in the electricity consumption curve, location, and electricity cost. The remaining input data that remain unchanged are shown in the Table 3.

Table 3. Input data that remained unchanged during all tests.

Number of PV modules	9
Total unrestricted battery capacity (kWh)	5.04
Rotation of PV modules (azimuth)	180
Placement location of PV modules	Pitched roof
PV module price (Wp)	0.65

Household location was the same for all tested households as we had anonymized data available where the sites were not listed. The Table 4 shows two locations. One was used only for Irish households, the other for Texas.

Table 4. Arbitrary set geographical coordinates of tested household.

City	Dublin	Austin
Latitude	53.359151	30.268076
Longitude	−6.280032	−97.742220
Altitude (m a.s.l.)	40	162

4.1 Dublin, Ireland

The price for electricity was set to €0.1734 per kWh [7]. We tested 709 households. We present two households - with the highest and lowest net savings (Fig. 2 and Table 5).

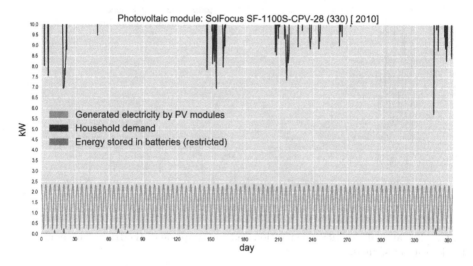

Fig. 2. The annual electricity generation and consumption of the Irish household with the highest net savings.

Irish household with the highest net savings had a huge electricity consumption and therefore almost all the generated electricity from the PV modules was used to cover the household consumption. Only about 0.35 kWh of electricity was not used. The results also show that it would not even be necessary to use the batteries for this household. The batteries could not even be charged when almost all the electricity generated from the PV modules was used to cover the household consumption (Fig. 3 and Table 6).

Irish household with the lowest net savings had very small electricity consumption and therefore the huge amount of electricity generated by PV modules remained untapped. As a result, net savings were very low and the investment into PV system would not pay off for this household.

Table 5. Irish household with the highest net savings.

Recommended PV module	SolFocus SF-1100S-CPV-28 (330) [2010]
Generated el. from PV modules (1 yr)	5851.32 kWh
Household demand (1 yr)	71777.96 kWh
El. consump. cov. by PV modules (1 yr)	5834.61 kWh
El. consump. cov. from batteries (1 yr)	16.36 kWh
Max. output power from PV modules	2.97 kW
Total costs (20 yrs)	€ 6344.98
Net savings (20 yrs)	**€ 14015.9**

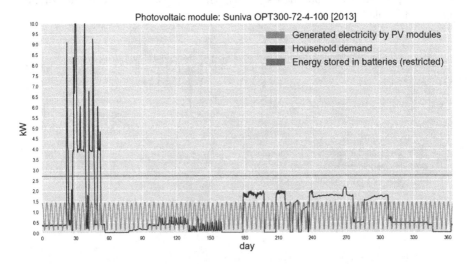

Fig. 3. The annual electricity generation and consumption of the Irish household with the lowest net savings.

Table 6. Irish household with the lowest net savings.

Recommended PV module	Suniva OPT300-72-4-100 [2013]
Generated el. from PV modules (1 yr)	5157.79 kWh
Household demand (1 yr)	1140.56 kWh
El. consump. cov. by PV modules (1 yr)	258.42 kWh
El. consump. cov. from batteries (1 yr)	115.35 kWh
Total savings (20 yrs)	€ 1300.65
Total costs (20 yrs)	€ 5342.86
Net savings (20 yrs)	**€ −4042.21**

4.2   Austin, Texas

The price for electricity has been converted from dollars and set to € 0.1016 per kWh [2]. We tested 115 households which were on the Pekan Street. Just as for Irish households, we also present Texas households with the highest and lowest net savings (Fig. 4 and Table 7).

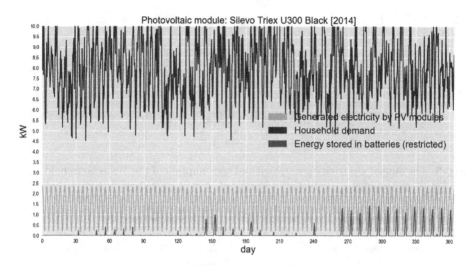

Fig. 4. The annual electricity generation and consumption of the Texas household with the highest net savings.

Table 7. Texas household with the highest net savings.

Recommended PV module	Silevo Triex U300 Black [2014]
Generated el. from PV modules (1 yr)	6590.42 kWh
Household demand (1 yr)	33471.83 kWh
El. consump. cov. by PV modules (1 yr)	6421.59 kWh
El. consump. cov. from batteries (1 yr)	129.76 kWh
Total savings (20 yrs)	€ 13358.11
Total costs (20 yrs)	€ 6117.5
Net savings (20 yrs)	**€ 7240.61**

Texas household with the highest net savings had a huge electricity consumption and therefore almost all the generated electricity from the PV modules was used to cover the household consumption. Only approximately 39.07 kWh of electricity was not used. Using batteries for this household would not be an economically viable option (Fig. 5 and Table 8).

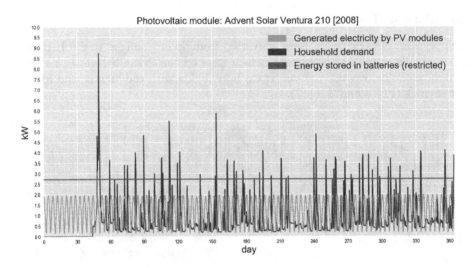

Fig. 5. The annual electricity generation and consumption of the Texas household with the lowest net savings.

Table 8. Texas household with the lowest net savings.

Recommended PV module	Advent Solar Ventura 210 [2008]
Generated el. from PV modules (1 yr)	4533.53 kWh
Household demand (1 yr)	848.21 kWh
El. consump. cov. by PV modules (1 yr)	236.26 kWh
El. consump. cov. from batteries (1 yr)	115.34 kWh
Total savings (20 yrs)	€ 716.85
Total costs (20 yrs)	€ 4834.48
Net savings (20 yrs)	€ −4117.63

Texas household with the lowest net savings had very small electricity consumption and therefore the huge amount of electricity generated by PV modules remained untapped. Batteries in this case helped to increase net savings, but household electricity consumption was so low that the investment in the PV system would not pay off for this household.

4.3   Comparison with Brute Force

Experimentally, we verified the time complexity of our algorithm compared to brute force. The chart in the Fig. 6 shows the average of three measurements for four different sizes of PV module datasets.

For the main advantage of our algorithm, we consider the fact that with the size of the PV module dataset, the time complexity increases only minimally.

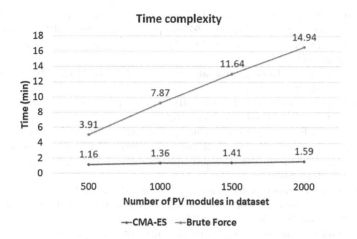

Fig. 6. Time complexity of algorithms due to the size of the PV module dataset.

5   Conclusion and Future Work

We implemented an algorithm that recommends a specific PV module from the database and calculates the return on investment, based on yearly average data. As the results shows, the size of the PV module database affects the time complexity of the calculation only minimally Fig. 6. We provide a free Android application for households[1]. We compared 709 Irish and 115 Texas households (extensive results are available in the GitLab repository (see Footnote 1)).

In the future, we plan to add an option to load profiles of electricity consumption for people without access to their historical data. We also plan to add optimization of battery capacity and day/night tariff for electricity pricing. As for the Android application, we plan to make a selection of a household location using Google Maps and also according to current device location.

References

1. IEEE Standards Coordinating Committee 21: IEEE guide for selecting, charging, testing, and evaluating lead-acid batteries used in stand-alone photovoltaic (PV) systems. IEEE Std 1361-2014 (Revision of IEEE Std 1361-2003), pp. 1–39, June 2014. https://doi.org/10.1109/IEEESTD.2014.6837414
2. Bounce Energy: Electricity Prices. https://www.bounceenergy.com/. Accessed 26 June 2018
3. ESolar: ESolar. https://www.esolar.sk/. Accessed 24 Mar 2018
4. Hansen, N., Ostermeier, A.: Completely derandomized self-adaptation in evolution strategies. Evol. Comput. **9**(2), 159–195 (2001)

---

[1] https://gitlab.com/DavidKubik/Photovoltaic_Module_Optimizer.

5. Khalilpour, K.R., Vassallo, A.: Technoeconomic parametric analysis of PV-battery systems. Renew. Energy **97**(Suppl. C), 757–768 (2016). https://doi.org/10.1016/j.renene.2016.06.010. http://www.sciencedirect.com/science/article/pii/S0960148116305225

6. National Renewable Energy Laboratory: System Advisor Mode. https://sam.nrel.gov/. Accessed 22 Feb 2018

7. Panda Power: Electricity Prices. https://www.pandapower.ie/plan/cashback-bundle-electricity/. Accessed 26 June 2018

8. ENF Solar: FLEXmax MPPT80 Charge Controller. https://www.enfsolar.com/pv/charge-controller-datasheet/2238?utm_source=ENF&utm_medium=charge_controller_list&utm_campaign=enquiry_product_directory&utm_content=21404. Accessed 24 Mar 2018

9. Wholesale Solar: Fullriver DC210-12 AGM Sealed 12V 210Ah Battery. https://www.wholesalesolar.com/9949469/fullriver/batteries/fullriver-dc210-12-agm-sealed-12v-210ah-battery. Accessed 24 Mar 2018

10. Google Project Sunroof: Project Sunroof. https://www.google.com/get/sunroof. Accessed 22 Feb 2018

11. Weniger, J., Tjaden, T., Quaschning, V.: Sizing of residential PV battery systems. Energy Procedia **46**(Suppl. C), 78–87 (2014). https://doi.org/10.1016/j.egypro.2014.01.160. http://www.sciencedirect.com/science/article/pii/S1876610214001763. 8th International Renewable Energy Storage Conference and Exhibition (IRES 2013)

Fused Lasso Dimensionality Reduction of Highly Correlated NWP Features

Alejandro Catalina$^{(\boxtimes)}$, Carlos M. Alaíz, and José R. Dorronsoro

Departamento de Ingeniería Informática and Instituto de Ingeniería del
Conocimiento, Universidad Autónoma de Madrid, Madrid, Spain
Alejandro.catalina@uam.es

Abstract. Two problems when using Numerical Weather Prediction
features in Machine Learning are the high dimensionality inherent to
the current high-resolution models, and the high correlation of the features, which can affect the performance of learning machines as Multilayer Perceptron (MLP). In this work we propose to reduce the dimension
of the problem using a supervised Fused Lasso model, which generates
meta-features corresponding to the average of the groups with constant
coefficients. The Fused Lasso problem is defined in terms of the feature
correlation graph and tries to retain features with the stronger connections. As shown experimentally, training the models over the correlation
graph-based reduced dataset allows to decrease the overall computational
time while preserving almost the same error in the case of Support Vector Regressors and even improving the error of the MLPs, if the original
dimension is high.

1 Introduction

Big data problems often involve very high dimensional data with large correlations among them. This implies that model building will be quite costly and,
moreover, that high feature correlations may make model training difficult. A
particularly clear and important example is given by the Numerical Weather
Prediction (NWP) features to be used in renewable energy predictions. These
are computed over a grid and it is clear that a variable such as, say, the U
component of wind will show high correlations at nearby grid points. Moreover,
grid resolution is steadily increasing: that of the European Center for Medium
Weather Forecasts (ECMWF) is currently about 0.1° but for mesoscale systems
may be quite lower. If we add to this the need for predictions over a relatively
large area, pattern dimension may shoot up while the large correlations will still
be there.

Some form of dimensionality reduction is thus highly advisable and it is quite
natural to use for this the information in the sample correlation matrix. As it is
well known, this is done for instance in Principal Component Analysis (PCA),
where the selected orthogonal eigenvectors are used to build new features which
are then used afterwards. PCA is an example of a filter approach, in the sense

© Springer Nature Switzerland AG 2018
W. L. Woon et al. (Eds.): DARE 2018, LNAI 11325, pp. 13–26, 2018.
https://doi.org/10.1007/978-3-030-04303-2_2

that no predictive model is built. The alternative are sparsity inducing wrapper methods, of which the Lasso is a well-known example. Lasso models incorporate variance information in an implicit way and their handling of highly correlated variables may lead to an unstable feature selection in the sense that it tends to select somewhat randomly only one of them; see for instance [8, Subsect. 2.1]. Several ideas have been proposed to alleviate this, such as ElasticNet, either direct [15] or pairwise [12], Trace Lasso [8], Network Lasso [9], or, when some group structure among variables is known, Group Lasso [14].

Recently, ad-hoc regularizers that precisely aim to handle the high correlation situation have been proposed; see [3,7] and, particularly, [10,11], where a correlation matrix regularizer is defined by a concrete generalized Lasso problem over the complete graph determined by the correlation matrix. More precisely, the computational complexity of generalized Lasso may be quite high and because of this, when dealing with fMRI data, the authors in [11] choose to work with a skeleton graph given by a Maximum Spanning Tree (MST) over the complete graph defined by the absolute values of the correlation matrix, solving the generalized Lasso over the resulting tree graph. In general, spanning trees will not be paths (i.e., degree one spanning trees), but since the underlying correlation graph is complete, one can get such a path simply by traversing the spanning tree by a depth first search (DFS).

While the goal of [11] is to provide theoretical guarantees for the linear least squares estimator when regularized by correlation, the goal in [10] is to simplify the denoising problem over general graphs by replacing the general graph based Fused Lasso problem (also referred sometimes as Total Variation denoising TV_G over the graph G)

$$\hat{\theta}_G = \arg \min_{\theta} \frac{1}{2} \|y - \theta\|_2^2 + \lambda \|\nabla_G \theta\|_1, \tag{1}$$

where y is an appropriate signal to be denoised, with a much simpler Fused Lasso problem [13] over a certain chain sub-graph of the original one built by computing first an MST and then traversing it by DFS. The key for this is the observation that the mean squared error of the signal denoised using (1) is about the same than that achieved when replacing G by the MST–DFS chain graph.

Now this chain graph is just a path in G and, as such, when applied to a feature graph, it induces an obvious ordering in the original features; we can thus simply reorder the feature vertices ending up with a standard Fused Lasso (FL) problem. Obviously, FL problems are easier to solve and, in fact, the well-known Taut String algorithm [5] provides an exact solution in contrast with the iterative one that Generalized Lasso requires.

While the discussion in [10] centers on denoising, it can obviously be extended to linear least squares problems regularized by the correlation graph total variation. Linear models are likely to produce errors in NWP-based renewable energy problems larger than those achievable by more powerful models such as Deep Neural Networks (DNNs) [6] or Support Vector Regressors (SVRs) [4]. On the other hand, DNNs may be greatly affected by feature correlations (SVRs appear to be more resilient). Thus, while intriguing, least squares, correlation regularized models are likely not to be a match for DNNs or SVRs. However, an important advantage of FL is that it groups the path-ordered features by those having the same coefficients in the FL model. Moreover, the combined effect of such a grouping is thus the same of that of a single variable that averages them, and this suggests to use these group averages as a simple way to achieve supervised dimensionality reduction.

In this work we are going to explore the effect and advantages of such an approach to dimensionality reduction. More precisely:

1. We will describe formally and algorithmically the dimensionality reduction procedure we have just introduced.
2. We will compare over several renewable energy regression problems the results of MLP and SVR models built over an entire set of original NWP features with those of the same MLP and SVR models built now over the reduced features obtained by the proposed procedure.
3. We will perform the same comparisons when the same dimensionality reduction procedure is applied over the natural line-by-line order of features over the NWP grid.

The rest of the paper is organized as follows. In Sect. 2 we will briefly review the ideas in [10,11] most relevant for us. We describe our dimensionality reduction approach in Sect. 3 and apply it in Sect. 4 to wind energy prediction at an individual farm and also over Peninsular Spain. The paper ends with a short discussion and pointers to further work.

2   Covariance Graph-Based Regularization

As mentioned, high dimensional problems have often associated highly correlated features, which makes their correct handling crucial not only for the obvious case of linear models but also for others such as multilayer perceptrons with partially linear components. A natural idea is thus to exploit the complete graph structure determined by the correlation matrix $C = (c_{j,k})$ to alleviate undesired correlation effects. This is the approach followed in [11] where a linear model $\hat{y} = x \cdot w$ is sought by minimizing the following cost function:

$$\frac{1}{N} \|y - Xw\|_2^2 + \lambda_S \sum_{j,k} |c_{j,k}| \left(w_j - s_{j,k} w_k\right)^2$$

$$+ \lambda_{TV} \sum_{j,k} |c_{j,k}|^{1/2} |w_j - s_{j,k} w_k| + \lambda_1 \|w\|^2, \tag{2}$$

with $s_{j,k} = \text{sign}(c_{j,k})$ and λ_S, λ_{TV} and λ_1 are the penalty factors associated to the Laplacian smoothing, graph Total Variation (TV_G) and sparsity regularizers, respectively. Since in our case we are primarily interested in simplifying the feature structure, the cost function we want to minimize is

$$\frac{1}{N} \|y - Xw\|_2^2 + \lambda \sum_{j,k} |c_{j,k}|^{1/2} |w_j - s_{j,k} w_k|. \tag{3}$$

Given the computational complexity imposed by a full TV regularizer on large graphs, in a large dimensional fMRI problem in [11, Sect. 5.1] the full graph is replaced by a skeleton chain graph S obtained by finding first a Maximum Spanning Tree (MST) T on the weighted graph defined by the matrix $|C|$ of absolute correlations and then by running DFS on T. Notice that the ordering in S is just a permutation P on $[1, \ldots, d]$ and if we reorder the original features by P and renumber them again as $[1, \ldots, d]$, the new problem is to minimize

$$\frac{1}{N} \|y - Xw\|_2^2 + \lambda \sum_{j=1}^{d-1} |c_{j,j+1}|^{1/2} |w_j - s_{j,j+1} w_{j+1}|, \tag{4}$$

that is, a standard FL problem [13].

No reasons other than computational convenience are given in [11] for this chain graph heuristic but recent results in [10] can be used to support it. In fact, notice first that the standard way of solving problems (3) and (4) is through the FISTA algorithm [2] where a gradient step is followed by the application of the proximal operator of the regularizer, namely the graph TV denoiser

$$\mathrm{prox}_{\mathrm{TV}_G}(w) = \arg \min_{\theta} \left\{ \sum_{j,k} |c_{j,k}|^{1/2} |\theta_j - s_{j,k} \theta_k| + \frac{1}{2} \|\theta - w\|^2 \right\} \tag{5}$$

for the full graph and the one-dimensional TV denoiser

$$\mathrm{prox}_{\mathrm{TV}}(w) = \arg \min_{\theta} \left\{ \sum_{1}^{d-1} |c_{j,j+1}|^{1/2} |\theta_j - s_{j,j+1} \theta_{j+1}| + \frac{1}{2} \|\theta - w\|^2 \right\} \tag{6}$$

for the chain graph. While (5) can only be solved approximately and in a costly iterative way, the Taut String algorithm [5] solves (6) exactly and with a linear cost. Moreover, while in principle the TV and FL proximals will yield different solutions, it turns out that, as shown in [10] for unweighted graphs, the recovery errors for properly tuned graph TV and DFS TV denoisers are similar and, in an appropriate sense, while seemingly simpler, the chain graph is among the hardest graphs for denoising bounded variation signals.

We will follow here a middle path, building a chain graph by first looking over an MST T on the complete graph with adjacency matrix C, running DFS on T to obtain a permutation P of the original features and, after reordering them according to P and renumbering them as $[1, \ldots, d]$, solving the standard FL problem

$$w^* = \arg \min_{w} \left\{ \frac{1}{N} \|y - Xw\|_2^2 + \lambda \sum_{j=1}^{d-1} |w_j - w_{j+1}| \right\}, \tag{7}$$

which is a simplification of (4) assuming $c_{j,j+1} \simeq 1$ and thus $s_{j,j+1} = 1$. Given our choice of the spanning tree, both assumptions are likely to be true for the initial edges of T although they are not guaranteed for all j.

Once we have solved (7) for an adequate λ, the w^* coefficients should have a piecewise constant structure, in which features that are highly correlated will have assigned the same value in their coefficients. This resulting grouping is actually equivalent to having a reduced feature set in which all the features in a group are averaged and their corresponding (constant) weights added. Given that linear models are usually poorer than stronger counterparts such as MLPs or SVRs, a natural way to exploit the smaller and less correlated features derived from this dimensionality reduction is to build upon them complex models such as MLPs, which may otherwise suffer when facing large number of highly correlated features, or to lower the computational costs of SVRs that may be more robust when facing correlated features. We describe the exact procedure next.

3   Correlation Graph-Based Dimensionality Reduction

Our proposal for correlation graph-based dimensionality begins by identifying a chain graph on the correlation graph whose vertex ordering induces a feature permutation where consecutive features often have a high positive correlation. We do so as follows:

- Compute the correlation matrix C, i.e., the adjacency matrix of the complete undirected correlation graph.
- Compute an MST T on C by applying Kruskal algorithm for minimal spanning trees on the adjacency matrix given by $I - C$. Notice that we expect T to have a large number of edges (k_i, k_{i+1}) with weights $1 - c_{k_i, k_{i+1}} \simeq 0$, i.e., $c_{k_i, k_{i+1}} \simeq 1$.
- Traverse T by DFS and reorder the features and renumber them in $[1, \ldots, d]$ according to the DFS permutation P.

In principle, DFS visits the vertices of T according to their positions on the adjacency list of each vertex. If these lists are short, i.e., if the maximum degree of T is small, we can expect essentially all of the list weights to be large and close to 1. We build next an optimal FL model on the permuted variables by:

- Computing the optimal FL penalty λ^* over a validation set using the Mean Absolute Error (MAE) as the validation criterion function.
- Solving the linear FL problem (7) for λ^* to get the optimal coefficients w^*.

This will produce a linear model which gives the same weight to all features included in the groups induced by FL. Finally we perform dimensionality reduction by:

- Grouping the variables in M subsets $G_m = \{x_{p_m}, \ldots, x_{p_m + N_m - 1}\}$, $1 \leq m \leq M$, where now x denotes the reordered and renumbered features, N_m is the number of features in G_m and $w^*_{p_m} = \ldots = w^*_{p_m + N_m - 1}$.
- Defining the reduced feature set $x_1^R, \ldots, x_M^R$ as

$$x_m^R = \frac{1}{|G_m|} \sum_{x_j \in G_m} x_j,$$

by averaging the values of the features that have the same weight in the linear FL model.

The reduced dimension will thus be the number of feature groups induced by FL. Notice that defining the weight vector w^R as $w_m^R = |G_m| w_{p_m}^*$, $1 \leq m \leq M$, the FL model defined by w^* on the initial x features gives the same predictions that the one defined by w_R^* on the x^R features. In any case, once the x_m^R have been computed, we will use them to build optimal MLP or SVR models.

As just presented, the proposal corresponds to a wrapper-type dimensionality reduction method for which there is an obvious trade-off between model accuracy, i.e., a small MAE achievable when the FL regularization penalty is light enough, and the final dimension, i.e., a small number of FL groups, which can be achieved by strong regularization. Both settings are determined by the FL penalty factor λ: a large λ will result in few groups but a possibly large error, something which is reversed with a small λ parameter.

Finally, the complexity of the proposed procedure can be analyzed as follows. Provided that the $d \times d$ covariance matrix C has been already computed, applying the correlation graph-based dimensionality reduction proposed here requires to build the MST with a cost of $\mathcal{O}(d^2 \log d)$, to traverse it with a cost $\mathcal{O}(d)$, and to solve a classical Fused Lasso problem, with a cost of $\mathcal{O}(d^2 k)$, with k the number of iterations of the FISTA algorithm; for the high-dimensional problems tackled here we can expect $k \ll d$. For comparison purposes, when C is known then PCA, the most commonly used dimensionality reduction method, has a cost $\mathcal{O}(d^3)$ if no assumptions about the number of reduced dimensions is made and a full eigenanalysis is needed.

4   Numerical Experiments

4.1   FL-Based Dimensionality Reduction

We will apply the proposed dimensionality reduction procedure to two wind energy prediction problems, that of the Sotavento wind farm in north-western Spain and that of the entire peninsular Spain using as features NWP forecasts of 10 variables over the points of a rectangular grid that contains the Iberian Peninsula. These variables are the U and V wind speed components at 10 and 100 m, and 2 m temperature and surface pressure; to these 6 variables we then add 10 and 100 m wind speed modules and the corresponding wind-turbine power computed using a generic wind-to-power conversion curve. The energy production in both problems is normalized as a percentage of the installed power.

For Peninsular Spain we will work with a rectangular grid with a lower left corner with coordinates $(35.5°, -9.5°)$ and upper right corner with coordinates $(44.0°, 4.5°)$, latitude and longitude respectively. Currently the spatial resolution of the ECMWF forecasts is of $0.1°$; for simplicity we will consider resolutions of $0.25°$ and $0.5°$, and will name these datasets REE$_{025}$ and REE$_{050}$, respectively. The number of grid points are thus 1,995 for the $0.25°$ resolution and 522 for the $0.5°$ one. Consequently, the number of variables, i.e., problem dimensions are 19,950 for the 0.25° resolution and 5,520 for the 0.5° (notice that a 0.125° grid would result in 77,970 features and a correlation graph with approximately 6×10^9 edges, rather large for graph algorithms). For the Sotavento experiment, we will consider a smaller rectangular sub-grid with lower left coordinates $(42.25°, -9.5°)$ and upper right coordinates $(44.0°, -6.0°)$ and only the 0.25° resolution. The resulting dataset, Sot_{025}, has a total of 120 grid points and thus 1,200 variables.

Once we have computed the chain graph feature orderings as described in the previous section, we determine the optimal FL λ penalty parameter by validation, using NWP and production data of 2013 for training purposes and those of 2014 as the validation set; we use the FL implementation proposed in [1]. We will examine logarithmically equispaced λ values 10^k with $-8 \leq k \leq 6$, i.e., 15 penalty values, and choose the optimum λ^* giving a smaller MAE in 2014.

As a sanity check we will also consider as baseline for comparison against our proposal a natural latitude–longitude ordering over each feature group which takes advantage of the natural block and grid structure of the NWP features. It thus ensures high correlations between consecutive features so, in principle, it should also give a sensible FL model (notice that, while present in other problems such as image processing, a grid structure ensuring high correlations will not be available in general problems). Its optimal λ is determined in the same way as for the chain graph ordering.

Table 1 shows the initial problem dimensions and those after the FL reductions over the MST–DFS (DFS for short) and latitude–longitude (LL for short) feature orderings; it also gives the optimal λ^* FL penalties. As it can be seen, DFS and LL give a considerable dimensionality reduction in all problems. Compared with the LL one, the DFS reduction is smaller on Sot_{025}, slightly larger on REE_{050} and much larger on REE_{025}. While this might seem surprising at first sight, observe that since the number of features in REE_{025} is four times that of REE_{050}, the corresponding dimensionality reduction should roughly be about 4 times larger. Moreover, one would expect the final dimensions to be similar; actually the REE_{025} one is about half that of REE_{050}. These facts seem to point to an intrinsic dimension between 200 and 400 for the NWP features needed to predict Peninsular Spain's overall wind energy.

Table 1. Dimensions of the original datasets and the DFS and LL reduced datasets, corresponding reduction ratios and optimal FL λ^* penalties.

	Dimensions			Ratios		λ^*	
	Full	DFS	LL	DFS	LL	DFS	LL
REE_{050}	5,520	340	391	16.24	14.12	10^{-3}	10^{-3}
REE_{025}	19,950	177	1,233	112.71	16.18	10^{-2}	10^{-2}
Sot_{025}	1,200	106	110	11.32	10.91	10^{-3}	10^{-3}

In fact, if we observe the λ–MAE and λ–sparsity evolution for the DFS model over REE_{050} in Fig. 1, we can see that after the optimum MAE is achieved at $\lambda^* = 10^{-3}$, the number of FL blocks grows quite rapidly. Possibly a more refined exploration of λ values around 10^{-3} would lead to a more precise regularization and, hence, dimensionality reduction. Of course, such an exploration might also result in a better $0.25°$ dimensionality reduction by the LL ordering. Figures 2 and 3 show a similar situation for the REE_{025} and Sot_{025} datasets.

(a) Evolution of the MAE.   (b) Evolution of the FL groups.

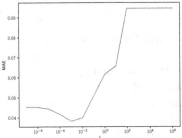
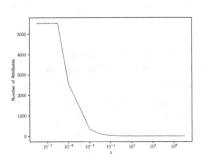

Fig. 1. Evolution with respect to λ of the validation MAE and the number of FL groups for REE_{050}.

(a) Evolution of the MAE.   (b) Evolution of the FL groups.

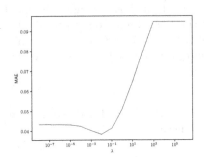
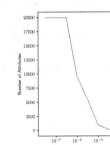

Fig. 2. Evolution with respect to λ of the validation MAE and the number of FL groups for REE_{025}.

Finally, in order to better understand the new variable structure, we show in Figs. 4, 5 and 6 the correlation heatmaps of the original features and of the optimally reduced features for REE_{050}, REE_{025} and Sot_{025}. The ordering for the

(a) Evolution of the MAE.   (b) Evolution of the FL groups.

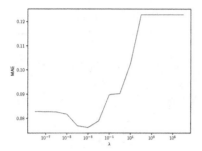
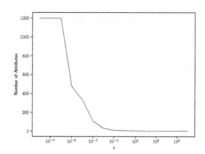

Fig. 3. Evolution with respect to λ of the validation MAE and the number of FL groups for Sot_{025}.

full features is done first by feature groups and then by the previous LL order, sliding first by longitude and afterwards by latitude; we start at the upper left corner and finish in the lower right one. The full feature maps show a clear block correlation structure, which is stronger for Sot_{025} (whose regional area is much smaller) and less so for REE_{050}. Another diagonal structure is also present on the reduced features but it is clearly weaker, particularly for REE_{050} and REE_{025}, which present a much narrower diagonal structure and weaker near diagonal correlations. This indicates that, while still far from independent, the correlation structure of the new features is clearly weaker.

(a) Original REE_{050}.   (b) Reduced REE_{050}.

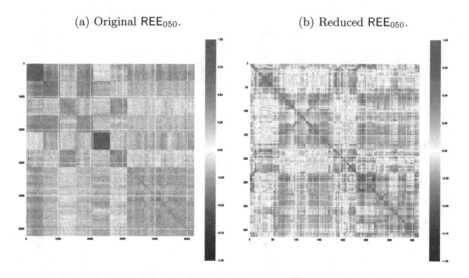

Fig. 4. Correlation heatmaps for the original and reduced REE_{050} datasets.

(a) Original REE$_{025}$.   (b) Reduced REE$_{025}$.

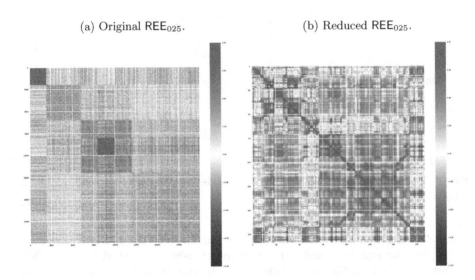

Fig. 5. Correlation heatmaps for the original and reduced REE$_{025}$ datasets.

(a) Original Sot$_{025}$.   (b) Reduced Sot$_{025}$.

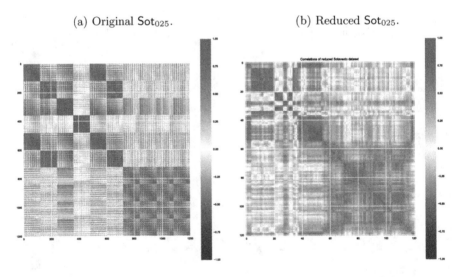

Fig. 6. Correlation heatmaps for the original and reduced Sot$_{025}$ datasets.

4.2   MLP and SVR Full and Reduced Models

We finally compare the test MAEs of MLP and Gaussian SVR models built over either the original x or the DFS or LL reduced x^R features. We consider MLPs with 4 layers and with 1,000 units each on REE and 100 units for Sotavento. Moreover, due to the inherent randomness of MLPs, we will report the average results of 5 MLPs trained from randomly different initial weights. In order to obtain the ℓ_2 regularization penalty α for MLP and the SVR C, γ and ϵ parameters we will perform grid searches with 2013 as the training set, 2014 as the validation one; the reported test results are for 2015. We use the standard scikit-learn implementations of feed-forward MLPs and SVRs.

Table 2. MAE test results (in percentage) for full and reduced MLP and SVR models.

	MLP			SVR		
	Full	DFS	LL	Full	DFS	LL
REE_{050}	2.76	3.00	2.94	2.54	2.64	2.77
REE_{025}	3.28	3.06	3.19	2.54	2.72	2.64
Sot_{025}	5.86	5.96	5.98	5.80	5.79	5.78

In Table 2 these test MAEs are compared for the full and the DFS and LL reduced MLP and SVR models at the three datasets REE_{050}, REE_{025} and Sot_{025}. As it can be seen, the DFS models are competitive in all cases but only best for the REE_{025} MLP and the Sot_{025} SVR. This is partially to be expected: the possible performance gains of DFS models in limiting feature collinearity may be neutered by a potential loss of information.

On the other hand, this dimensionality reduction–performance trade-off could be better balanced. For instance, our current choice of the λ penalty is done in terms of the FL validation MAE. FL models are rather weak and have it hard to take advantage of feature information. On the other hand, looking at Figs. 1, 2 and 3, it can be seen that choosing instead of the FL optimal λ^* the ones right to their left (namely, 10^{-4}, 10^{-3} and 10^{-4} respectively), the reduced dimensionality grows moderately but the corresponding features are likely to be richer and, thus, result in better MLP and SVR models. This option should be explored in further work.

Table 3. Hyperparameterization times (minutes) for full and reduced MLP and SVR models.

	MLP			SVR		
	Full	DFS	LL	Full	DFS	LL
REE_{050}	935.6	701.3	792.9	1,213.0	129.2	156.4
REE_{025}	3,079.0	652.6	1,152.4	4,443.8	150.2	344.2
Sot_{025}	75.6	34.6	31.3	368.2	135.4	110.5

Moreover, this MAE comparison has to be complemented with the hyperparameterization times reported in Table 3, which shows the reduced models to be much faster, with times roughly aligned with the dimension ratios of Table 1 in the case of SVRs and also with the number of weight ratios of the full and reduced MLPs.

5   Discussion and Conclusions

Very large dimensional and highly correlated features are the usual situation when applying Machine Learning models to predict renewable energy from NWP forecasts. Some ML models such as Gaussian SVRs are rather robust against collinearity, as they essentially center a Gaussian at each support vector and pattern interaction only comes through Euclidean distances. On the other hand, the initial feature processing by MLPs is done through matrix–vector products, more sensible to feature correlations. Many proposals have been made for dimensionality reduction, particularly PCA, but the cubic cost of computing the eigenanalysis of the possibly quite large covariance matrix makes them quite costly to use.

Here we have explored an alternative also based on the covariance matrix information. Our starting point is the recent proposal in [11], which proposes linear models penalized by the covariance graph total variation regularizers that are applied to fMRI data processing. But even if properly regularized, linear models are usually no match for MLPs and, specially, SVRs. So our goal here has been different and we have exploited covariance matrix information to derive features of a much smaller dimension but still effective for non-linear models.

To do so, following the ideas in [10] we have replaced the full covariance graph and, therefore, its quite costly TV regularizer, by a much simpler chain graph whose regularizer is just that of a one-dimensional Fused Lasso. The corresponding optimal linear models produce a grouping of the initial features giving the same weight to all inside such a group. It is thus natural to average the features of these groups, defining this way a new reduced feature set. As we have seen, models over the reduced features are always competitive and, in the Sotavento case, yield the best model, with much better training times in all cases.

While the preceding results have an exploratory nature, they show that the proposed approach has a clear potential and deserves to be further studied. For instance, we can replace the basic Fused Lasso problem (7) with the more precise problem (6) that should involve variables with higher absolute feature correlations and properly handles their signs. Moreover, and as mentioned, we can improve the trade-off between dimensionality reduction and model performance by selecting not the optimal FL penalty λ^* but the less demanding one just to their left in the λ vs MAE curves. This will certainly result in slightly worse FL models but, anyway, they are usually weak and, on the other hand, the new reduced features, while having a larger dimensionality, should also be richer and give better MLP and SVR models. Another possibility could be to consider a multi-objective Pareto procedure that yield jointly optimal MAE and sparsities.

Furthermore, and from an algorithmic point of view one could replace our DFS ordering by other graph possibilities such as, for instance, TSP algorithms with an appropriate complexity. And, finally, we point out that our sanity check exploration of the natural ordering of NWP variables by type groups and then by latitude and longitude shows that the DFS results have been comparable and often better. Of course a natural ordering such as the previous NWP one will not be available in other high dimensional, high correlation problems, while our feature reduction procedure may result in such cases in better models and, possibly, further insights on the problems involved. We are currently studying these and other related questions.

Acknowledgements. With partial support from Spain's grants TIN2016-76406-P and S2013/ICE-2845 CASI-CAM-CM. Work supported also by project FACIL–Ayudas Fundación BBVA a Equipos de Investigación Científica 2016, and the UAM–ADIC Chair for Data Science and Machine Learning. We thank Red Eléctrica de España for making available wind energy data and gratefully acknowledge the use of the facilities of Centro de Computación Científica (CCC) at UAM. We also thank the Agencia Española de Meteorología, AEMET, and the ECMWF for access to the MARS repository.

References

1. Barbero, A., Sra, S.: Fast newton-type methods for total variation regularization. In: Proceedings of the 28th International Conference on Machine Learning (ICML-11), pp. 313–320. Citeseer (2011)
2. Beck, A., Teboulle, M.: A fast iterative shrinkage-thresholding algorithm for linear inverse problems. SIAM J. Imaging Sci. **2**(1), 183–202 (2009)
3. Bühlmann, P., Rütimann, P., van de Geer, S., Zhang, C.H.: Correlated variables in regression: clustering and sparse estimation. J. Stat. Plan. Inference **143**(11), 1835–1858 (2013)
4. Catalina, A., Dorronsoro, J.R.: NWP ensembles for wind energy uncertainty estimates. In: Woon, W.L., Aung, Z., Kramer, O., Madnick, S. (eds.) DARE 2017. LNCS (LNAI), vol. 10691, pp. 121–132. Springer, Cham (2017). https://doi.org/10.1007/978-3-319-71643-5_11
5. Condat, L.: A direct algorithm for 1-D total variation denoising. IEEE Signal Process. Lett. **20**(11), 1054–1057 (2013)
6. Díaz, D., Torres, A., Dorronsoro, J.R.: Deep neural networks for wind energy prediction. In: Rojas, I., Joya, G., Catala, A. (eds.) IWANN 2015. LNCS, vol. 9094, pp. 430–443. Springer, Cham (2015). https://doi.org/10.1007/978-3-319-19258-1_36
7. Figueiredo, M., Nowak, R.: Ordered weighted l1 regularized regression with strongly correlated covariates: theoretical aspects. In: Artificial Intelligence and Statistics, pp. 930–938 (2016)
8. Grave, E., Obozinski, G., Bach, F.: Trace lasso: a trace norm regularization for correlated designs. In: Proceedings of the 24th International Conference on Neural Information Processing Systems, NIPS 2011. pp. 2187–2195 (2011)
9. Hallac, D., Leskovec, J., Boyd, S.: Network lasso: clustering and optimization in large graphs. In: Proceedings of the 21th ACM SIGKDD International Conference on Knowledge Discovery and Data Mining, KDD 2015, pp. 387–396 (2015)

10. Hernan Madrid Padilla, O., Scott, J.G., Sharpnack, J., Tibshirani, R.J.: The DFS Fused Lasso: Linear-Time Denoising over General Graphs. ArXiv e-prints, August 2016
11. Li, Y., Raskutti, G., Willett, R.: Graph-based regularization for regression problems with highly-correlated designs. ArXiv e-prints, March 2018
12. Lorbert, A., Eis, D., Kostina, V., Blei, D., Ramadge, P.: Exploiting covariate similarity in sparse regression via the pairwise elastic net. In: Teh, Y.W., Titterington, M. (eds.) Proceedings of the Thirteenth International Conference on Artificial Intelligence and Statistics. Proceedings of Machine Learning Research, PMLR, Chia Laguna Resort, Sardinia, Italy, vol. 9, pp. 477–484, 13–15 May 2010
13. Tibshirani, R., Saunders, M., Rosset, S., Zhu, J., Knight, K.: Sparsity and smoothness via the fused lasso. J. R. Stat. Soc. Ser. B (Stat. Methodol.) 67(1), 91–108 (2005)
14. Yuan, M., Lin, Y.: Model selection and estimation in regression with grouped variables. J. R. Stat. Soc. Ser. B (Stat. Methodol.) 68(1), 49–67 (2006)
15. Zou, H., Hastie, T.: Regularization and variable selection via the elastic net. J. R. Stat. Soc. Ser. B (Stat. Methodol.) 67(2), 301–320 (2005)

Sampling Strategies for Representative Time Series in Load Flow Calculations

Janosch Henze[(✉)], Stephan Kutzner, and Bernhard Sick

Intelligent Embedded Systems, University of Kassel, Kassel, Germany
{jhenze,kutzner.stephan,bsick}@uni-kassel.de

Abstract. Power system analysis algorithms increasingly use time series with a high temporal resolution to assess operational and planning aspects of the power grid. By using time series with high temporal resolution, information is getting more detailed, but at the same time, the computational costs of the algorithms increase. With the help of our algorithm, we create representative time series that have similar characteristics to the original time series. With the help of these representative time series, it is possible to reduce the computational cost of power system analysis algorithms having nearly the same results as with the original time series. In this work, we improve our previous algorithm with the help of specialized sampling strategies. Furthermore, we provide a new method to compare power analysis results achieved with the representative time series to the original time series.

Keywords: Representative time series · Feature extraction
Sampling · Load flow · Grid expansion planning · State estimation

1 Introduction

The data generation rate in a power grid increases with every new component in the grid. Especially, in the wake of smart grids as the power grid changes from a centralized generation to a decentralized generation, more data are generated. Due to this increasing amount of data and their increasing temporal and spatial resolution, we need specialized algorithms that help us to be able to perform calculations such as power grid planning, prediction of future energy production and consumption, and worst-case scenario analysis. One possibility is to compress the time series, e.g., by resampling, to obtain smaller time series with similar characteristics.

Our previous work tried tackling the problem of the increasing temporal and spatial resolution of data. The algorithm we proposed in [1] does this by creating representative time series. A representative time series is a time series which is shorter in length than the original input time series, yet maintains a similar amount of information.

The contribution of this article is the extension of our algorithm by additional sampling strategies, and by the capability of taking segmentwise correlation into

© Springer Nature Switzerland AG 2018
W. L. Woon et al. (Eds.): DARE 2018, LNAI 11325, pp. 27–48, 2018.
https://doi.org/10.1007/978-3-030-04303-2_3

account. By using different sampling strategies, the algorithm is even more capable of producing representative time series for different power system analyses. With the help of the capability to maintain segmentwise correlation, we can improve our previous results even more. Furthermore, we introduce a different method to compare the representative time series with the original time series.

Therefore, this article starts by looking at related work in Sect. 2. Afterward, we give a short overview of the improved algorithm and outline some flaws in the preceding article in Sect. 3. Section 4 explains the experiments and presents our new results. The article finishes with a conclusion and an overview of possible future work in Sect. 5.

2   Related Work

In our previous work [1] we presented a novel approach for finding representative time series with less amount of data than the original time series. We did this by identifying similar segments based on the features of a segment. In load flow applications such as grid expansion planning, worst-case scenario determination, or probabilistic load flow, we need to identify and represent the critical, task-specific time series segments of the original time series. By using different sampling strategies, we can select those features that create the best representative time series for such a task. Task-specific sampling for time series is used in different research areas, including energy informatics. Especially in probabilistic load flow calculation, sampling methods are used to create reliable results.

As mentioned in our previous work [1], the creation of representative time series for application in energy-related topics is an active field of research. Several researchers, e.g., [2–5], use feature extraction and clustering methods to create representative time series, which helps them to improve their results on energy-related questions. Other work focuses on more general questions, e.g., Merrick [6], on finding the optimal amount of data that preserves temporal variability to capture critical statistical features.

The focus of this article is on the sampling from clusters, especially in the field of load/power flow calculations. The most prominent applications of algorithms in this area is probabilistic load flow (PLF). PLF often uses Monte Carlo sampling or specializations of Monte Carlo Sampling, to obtain a distribution of possible results of a load flow calculation.

The authors of [7] use Latin supercube sampling (LSS) to obtain samples for probabilistic load flow calculation. The LSS allows them to consider the correlation of the input variables. They compared their sampling strategy to Latin hypercube and simple random sampling and showed that LSS is better in capturing the distribution of the output variables. They compared their results by using a bin-by-bin histogram comparison of the results. The authors of [8] use a similar approach, they use a Latin Hypercube Sampling method to incorporate photovoltaic generation into probabilistic load flow calculation.

Mitrah and Singh [9] use generation states, tie-line states, and load state scenarios from a pool of different states in a Monte Carlo simulation to solve DC Load flow for multi-area reliability evaluation.

Furthermore, the authors of [10] give an overview on the topic of load profiling and its application. Their focus is on different clustering methods that can be used to determine similar time series or time series segments.

Fischer et al. [11] use a bottom-up approach to generate load profiles of German households using an approach in which they sample short load time series of household appliances based on customer profiles derived from socio-economic factors, such as the family situation, housing types or age.

3   Outlier Based Sampling

We shortly explain the existing method for obtaining representative time series. Then we specify how we modify the sampling step to allow for task-specific sampling of representative segments within the feature space. Afterward, we explain each sampling method.

3.1   Generating Representative Time Series

Figure 1 shows our method to generate representative time series. It consists of six steps: preprocessing, segmentation, feature extraction, clustering, sampling, and postprocessing.

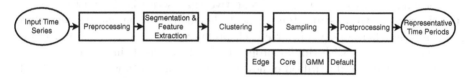

Fig. 1. A flow chart of the algorithm for the generation of representative time series using feature vectors with different sampling strategies. Shown are Edge, Core, GMM, and the default sampling method.

The method takes an equidistant time series $X = \{x_1, x_2, \ldots, x_T\}$ as input, where T is the length of the time series, and x_i are the data points for time step i. During the preprocessing step, we remove trends and seasonal influences by fitting a polynomial to a subsample of the original time series X. After that, we apply z-score normalization to make sure the preprocessed time series has a mean of 0 and a standard deviation of 1.

The preprocessed time series is then segmented into segments of equal length k: $Y = \{X_{(0,k)}, X_{(k+1,2 \cdot k)}, \ldots, X_{(T-k+1,T)}\}$. Afterward, for each segment features are extracted using either the sliding window feature extraction (SWFE) or the fast feature extraction (FFE) technique. The sliding window feature extraction yields features such as minimum and maximum values, median, mean, and variance of the time series in the sliding window. Therefore creating a feature vector $F_j = [\mathtt{min}(X_j), \mathtt{max}(X_j), \mathtt{median}(X_j), \mathtt{mean}(X_j), \mathtt{var}(X_j)]$ for the jth segment. In addition to the feature vector, we also store the starting index of the x_i of the segment.

The fast feature extraction uses the coefficients of orthogonal polynomials as features. A linear combination of polynomials with an orthogonal base is fitted in the sliding window under observation. It is also possible to combine both feature extraction algorithms and select the most relevant features, e.g., by using principal component analysis or any other feature selection technique.

Next, we have to find similar segments by clustering the feature vectors F. The clustering algorithm can be chosen freely. In our experiments, we selected a density-based clustering algorithm, which clusters data points based on their connectedness to nearby other data points. Furthermore, density-based clustering algorithms allow determining outliers, i.e., those data points that are not similar to other data points.

In the sampling step, we use the clusters to sample feature vectors. With the help of the stored information about the segment index, we convert the feature vector to the respective segment. By concatenating the sampled segments, we create a representative time series. In Sect. 3.2 we present four methods to select the samples for the creation of representative time series. Furthermore, we can select some outliers, detected by the density-based clustering, to be included in the representative time series. The outliers are all sampled with the same probability and sampled independently of the selected sampling strategy.

The final step to obtain a representative time series is the postprocessing step. During postprocessing, the segments belonging to the feature vector are recovered. Furthermore, we inverse each preprocessing step to receive a representative time series with similar properties as the original time series.

Figure 2 gives an illustration of how the reconstructed time series is created. Figure 2(a) shows a given input time series Y of the length $len(Y) = 31546000$, the number of seconds in a year. (b) visualizes the labeled data points obtained after preprocessing, segmentation, fast feature extraction, reduction of the data dimensionality using a principal component analysis as well as density-based clustering. (c) shows the related segments of the labeled data points, which presents the original time series after removing the yearly trend, applying z-score. (d) shows the reconstructed time series using 100 data points of (b) using core sampling.

3.2 Sampling Strategies

To obtain better results for load flow calculation, we adjusted the sampling strategy from a random sampling, i.e., the default method, to more specific sampling strategies, which allow picking the essential samples from our clusters.

For this purpose, we look at three different sampling strategies as shown in Fig. 3, core sampling, edge sampling, and GMM sampling. The GMM sampling should be similar to our default sampling strategy, whereas core and edge should be capable of obtaining better results in the extreme cases of the result

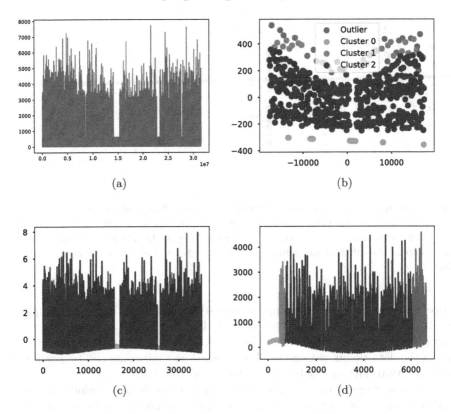

(a)   (b)

(c)   (d)

Fig. 2. An example showing the steps of the reconstruction process for one load flow time series. (a) shows the input time series. (b) reduction of the data dimensionality using pca and the resulting labels. (c) shows the segments in the original time series with applied zscore and yearly trend removal. (d) shows the representative time series with the according cluster information.

distribution. We compare the three strategies to our default sampling strategy from the preceding article.

Default: The default sampling strategy is based on the membership strength of a data point to a cluster. In density-based clustering algorithms, such as HDBSCAN [12], it is possible to determine the strength of membership to a cluster, which is normalized between 0 and 1.

We first create an ordered list of the data points. The list is ordered from high membership strength to low membership strength. We then generate a random number between 0 and 1. The first data point with a membership strength higher than the random number is selected from the ordered list and removed from the list. The process gets repeated until the desired amount of data points for the creation of the representative time series is collected.

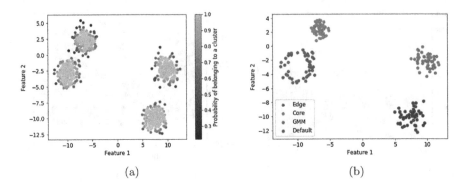

Fig. 3. This figure shows the result of the different sampling strategies. Each data point is a representation of a time series segment in the feature space. In (a) four different clusters with their respective probability of belonging to this cluster are shown. In (b) the results of the different sampling strategies are shown. Each strategy applied to one cluster.

Core Sampling: The core sampling chooses a quite intuitive way to improve the default sampling method. The core sampling only selects those samples with high connectivity, i.e., the core samples of the cluster. The core sampling strategy orders the data points within the cluster based on their connectivity, from high connectivity to low, and only selects the first n samples.

The advantage of this method is that only those samples which are most similar to each other get selected. Therefore, the method ignores the variance of the cluster and selects more similar time series segments for the creation of the representative time series. Therefore creating a representative time series which represents the core behavior of the original time series and ignores deviation from this behavior.

Edge Sampling: In contrast to the core sampling, the edge sampling only selects the edges of a cluster. Therefore it orders the cluster data points from low connectivity to high connectivity. The edge sampling strategy, as well as the others, select the first n elements for each cluster. Therefore, mainly covers the edges of the cluster.

This strategy will increase the variance of the resulting representative time series. Each of the segments should still look similar to other members of the same cluster, but the overall variety of segments should be higher. With the edge sampling strategy, the representative time series might not be as appropriate, but it will have more of the border cases of the original time series as components.

GMM Sampling: The GMM Sampling first fits a Gaussian Mixture Model to the cluster. In a second step, we draw n samples from the resulting distribution. This strategy will allow rebuilding the original time series similar to the default strategy. The main difference is that instead of taking the connectivity of the cluster into account as above, the samples are drawn based on their modeled density.

4 Evaluation

In our evaluation, we do not compare a representative time series to the original time series. We compare load flow results calculated with the representative time series and the original time series.

We perform two experiments. The first experiment is the same as in the previous article. In the second experiment, we segment the time series in such a way that we maintain the correlation of the time series within a segment, as explained earlier in Sect. 3.1. Therefore, allowing us to compare our new sampling strategies with the old results and furthermore, allows us to show how our sampling strategies allow for modeling the important areas for information about the power grid. The whole evaluation process is shown in Fig. 4.

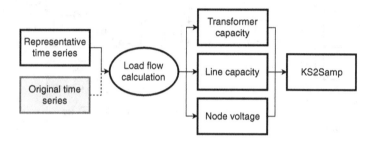

Fig. 4. Overview of the validation process.

Figure 4 shows that in comparison to our validation in [1], we replaced the Fisher permutation test with the Kolmogorov Smirnov two-sample test (KS2samp). The KS2samp test allows for comparing two distributions by their differences in the empirical cumulative density function (ECDF), as explained in more detail in Sect. 4.1.

For our evaluation, we created 100 representative load time series from their respective original time series. For each sampling method, a set of representative time series was generated, resulting in 4×100 representative time series. The original time series have a length of one year and a resolution of 15 m, which amounts to 35040 data points. The representative time series have a length of 6180 data points for the FFE method and 5700 for the SWFE method.

The representative time series and the original time series were used separately in a load flow calculation. In the first experiment we used a Kerber network as shown in Fig. 5 [13]. In the second experiment we used a network modelled for low voltage test scenarios from [14] as shown in Fig. 6. For each load in the network, a load had been assigned for each time step. In the first experiment, all loads in the power grid had the same load connected. In the second experiment, each load in the power grid had a different load connected. For the load flow calculation, we used pandapower as a load flow solver [15]. Within pandapower, we used the standard set of parameters for the calculation.

Furthermore, the second experiment takes segmentwise correlation of the original time series into account. For both experiments, the results of the load flow calculation for the representative time series and the original time series, i.e., the transformer capacity, the line capacity, and the node voltage, were compared using the KS2samp.

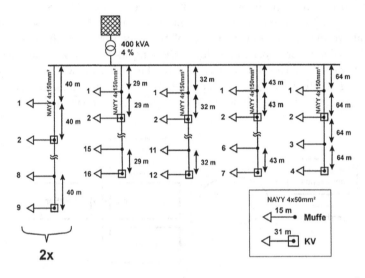

Fig. 5. Generic distribution grid used for the validation, as shown in [13].

4.1 Kolmogorov Smirnov Two Sample Test

The Kolmogorov Smirnov two sample test (KS2samp) compares two one-dimensional time series, $A = (a_1, a_2, ..., a_n)$ and $B = (b_1, b_2, ..., b_m)$. Whereas, their length n and m respectively do not have to be the same. The KS2samp allows to calculate a distance even independent of the underlying distribution and length [16].

The KS2samp first calculates the empirical cumulative distribution function (ECDF) for both time series, $F_{A,n}(x)$ and $F_{B,m}(x)$. Afterwards, the maximum vertical distance between those two functions is calculated by:

$$D_{A,B} = \sup_x |F_{A,n}(x) - F_{B,m}(x)| \tag{1}$$

$D_{A,B}$ can be interpreted as a distance measure describing how likely both time series originate from the same source. Low values of $D_{A,B}$ describe a high likelihood of originating from the same distribution and large values describe a low likelihood of originating from the same distribution.

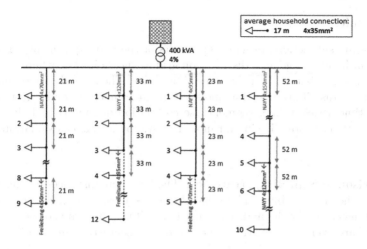

Fig. 6. Low voltage grid, modelled to represent a village in germany, as shown in [14].

In our evaluation, we compare the values of $D_{n,m}$ for the different methods to create a representative time series.

The D of the KS2samp can be interpreted as a distance between two time series, taking their underlying distributions into account. The Fisher test, on the other hand, is only able to reject the null hypothesis, based on the calculated p-value. P-value of statistical tests cannot be interpreted as a distance measure.

Exemplary results of the KS2samp can be seen in Fig. 7. In Fig. 7(a) a good representative time series is shown, with $D \approx 0.162$. In Fig. 7(b) a bad representative time series is shown with $D \approx 0.380$. The gap near the $1.0052pu$ indicates that such an representative time series does not perform the same as the original time series.

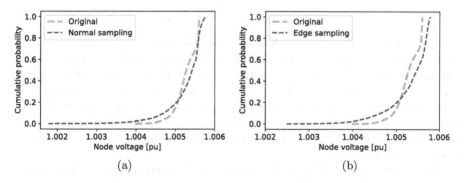

Fig. 7. A visualization of the empirical cumulative distribution functions (ECDF) for the load flow calculation of the original time series compared to the load flow calculation of the reconstructed time series using (a) SWFE method with GMM sampling and (b) SWFE method with edge sampling.

4.2 Results

The following subsections provide the results of the two experiments. The first experiment handles a generic distribution grid as shown in [13], where each load in the power grid is based on the same time series. The second experiment uses a low voltage distribution grid as modeled in [14], where all loads are based on different time series, taking segmentwise correlation of the original time series into account, therefore allowing for more realistic performance evaluation of the algorithm.

Comparison of Sampling Strategies: The KS2samp distances for this experiment are shown in Table 1. The table shows the mean and the corresponding standard deviation for the node capacity, line voltage, and transformer capacity. All values are averaged over the 100 created representative time series.

Table 1. KS2samp Distances for the first experiment. Displayed are mean values and standard deviations of the KS2samp distance D for the Line and Transformer Capacity, and the Node Voltage for each sampling strategy and segmentation method.

Segmentation	Strategy	Line capacity	Transformer capacity	Node voltage
FFE	Default	0.091(±0.045)	0.140(±0.061)	0.183(±0.096)
	GMM	0.096(±0.055)	0.143(±0.073)	0.165(±0.089)
	Edge	0.118(±0.053)	0.176(±0.069)	0.283(±0.116)
	Core	0.101(±0.047)	0.153(±0.066)	0.212(±0.117)
SWFE	Default	0.113(±0.052)	0.159(±0.067)	0.198(±0.103)
	GMM	0.097(±0.052)	0.143(±0.070)	0.166(±0.094)
	Edge	0.117(±0.052)	0.170(±0.678)	0.248(±0.116)
	Core	0.119(±0.053)	0.170(±0.069)	0.252(±0.118)

The results for the FFE method in Table 1 show that both Edge and Core sampling strategies perform worse than the Default and GMM strategy. The standard deviation of the GMM and the Edge sampling are among the highest. For the FFE method, the GMM strategy obtains the lowest KS2samp score regarding the node voltage. For the line and transformer capacity, the Default sampling method produces the lowest distances, closely followed by the GMM method with just some minor differences in the KS2samp distance which can be related to the averaging.

The results for the SWFE method show a similar picture. The GMM sampling again provides the lowest KS2samp value when comparing the representative time series to the original time series. The overall results for the SWFE are a bit worse than the results for the FFE method. More detailed results for the first experiment are given in Appendix A.

Maintain Segmentwise Correlation: The KS2samp distances for this experiments are shown in Table 2. The table shows the results of the load flow calculation. In this experiment, the different loads correspond to different households. Therefore, the results have not been averaged over all time series, as all time series are analyzed in the same load flow model. The values have been averaged over every line capacity and node voltage. As all loads have been distributed we only have one value to compare for the transformer capacity.

Table 2. KS2samp Distances for the second experiment. Displayed are the KS2samp distances D for the Line and Transformer Capacity, and the Node Voltage for each sampling strategy and segmentation method.

Segmentation	Strategy	Line capacity	Transformer capacity	Node voltage
FFE	Default	0.158(±0.053)	0.290	0.222(±0.025)
	GMM	0.174(±0.070)	0.349	0.296(±0.017)
	Edge	0.131(±0.047)	0.132	0.132(±0.006)
	Core	0.256(±0.083)	0.300	0.303(±0.010)
SWFE	Default	0.157(±0.057)	0.290	0.246(±0.016)
	GMM	0.172(±0.069)	0.354	0.292(±0.018)
	Edge	0.205(±0.065)	0.241	0.241(±0.015)
	Core	0.214(±0.079)	0.256	0.255(±0.016)

Table 2 shows that the FFE method has a lower KS2samp distance than the SWFE. The results for the different sampling strategies using the FFE for clustering show that the Edge strategy has overall the lowest scores. The Default and GMM strategy follow closely. The Core strategy which is similar to the Edge strategy obtains a high KS2samp distance.

For the SWFE segmentation method, the results are slightly different. The lowest KS2samp distances can be achieved for the Default strategy but only for the line voltage. For the remaining results, the Edge sampling strategy obtains the lowest KS2samp distances. Other good competitors for the line capacities are the Default and the GMM sampling strategies. More detailed results for the second experiment are given in Appendix B.

The second experiment allows evaluating task-specific sampling. As we have a more realistic setting, it is possible to have a closer look at the resulting empirical cumulative distribution functions (ECDF). We provide four images of the ECDF for the node voltage in Fig. 8. In Fig. 8(a), the whole ECDF for each sampling strategy is shown. The figure shows that most of the sampling strategies follow the ECDF of the original time series with deviations in the center. The subfigures (b) to (d) of Fig. 8 allow a more detailed analysis of the distribution. In Fig. 8(b), the lower left corner is displayed, and we can see that the Default and the GMM sampling strategies deviate a lot from the original. Furthermore, one can see that the Edge and Core strategies are more close to the original results, with

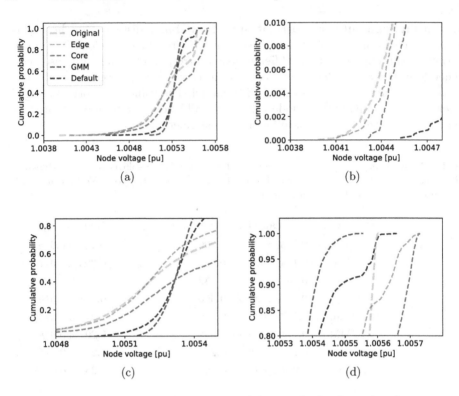

Fig. 8. This figure shows the ECDF of the load flow results for the node voltage at one node in (a). More close up views of the same plot are the lower left part shown in (b), the middle part shown in (c) and the upper right part shown in (d).

the Edge strategy being closest. For the center part of the ECDF in Fig. 8(c) we see that the sampling strategies deviate the most from the original time series. The closest ones are the Edge and the Core strategies, with the Edge strategy being the closest. In Fig. 8(d), the upper right part of the ECDF, we see that the results with the Edge and Default sampling strategies are closer to the results for the original time series than the Core and GMM strategies. These figures only depict one of the many results but give a visual impression of the capability of our algorithms.

5   Conclusion and Future Work

We presented three different sampling strategies for our algorithm that creates representative time series for load flow calculations. Furthermore, we introduced a more robust comparison method inspired by the Kolmogorov Smirnov two-sample test to compare the results of load flow calculations. We tested the strategies in two experiments, whereas the first experiment can be used to compare the results presented here to the results of our previous article. The second experiment allowed for an analysis of the sampling strategies in a more realistic setting and therefore in more detail. Both experiments showed that by introducing the different sampling strategies, we were able to improve our results.

During the evaluation, we showed that in comparison to the preceding article, the GMM strategy surpasses the Default sampling strategy. In the first experiment, a comparison test, the other sampling strategies, i.e., Core and Edge sampling did not show better results.

In the second experiment, we applied all strategies again in a more realistic setting, which showed that the Edge strategy in combination with the fast feature extraction method provides overall good results. Notably, in the extreme cases, the lower or upper end of the empirical cumulative probability density function, the Edge sampling produced values more close to the original ECDF than the GMM sampling strategy. The KS2samp values for the transformer capacity showed similar results.

In worst case scenario calculations, such as state estimation or specific grid planning scenarios, such information about the Edge cases is essential. Therefore, we propose to use the Edge sampling strategy in combination with the fast feature extraction method when looking at the stability or the maximum capability of the power grid.

Further research has to be conducted in the field of probabilistic load flow (PLF) calculations. In PLF, strategies such as Latin supercube sampling are used to select individual samples of a time series to obtain an underlying distribution of load flow results [7]. With the help of our representative time series, it might be possible to sample directly from the created feature space and to use the resulting time series segments as input for PLF calculations.

Acknowledgment. This work was created within the PrIME (03EK3536A) project and funded by BMBF: Deutsches Bundesministerium für Bildung und Forschung/German Federal Ministry of Education and Research.

Appendix A   Detailed Results - Univariate

See Tables 3, 4 and 5

Table 3. Load flow results for the line capacity for the first experiment.

	FFE								SWFE							
	Core		Default		Edge		GMM		Core		Default		Edge		GMM	
	Mean	Std	Mean	Std	Mean	Std	Mean	Std	Mean	Std	Mean	Std	Mean	Std	Mean	Std
0	0.380	0.022	0.340	0.023	0.375	0.020	0.302	0.020	0.365	0.022	0.323	0.023	0.385	0.022	0.298	0.020
1	0.377	0.022	0.338	0.022	0.372	0.019	0.300	0.020	0.363	0.021	0.321	0.022	0.382	0.022	0.295	0.019
2	0.367	0.021	0.327	0.021	0.361	0.019	0.290	0.019	0.354	0.020	0.310	0.021	0.372	0.022	0.285	0.018
3	0.339	0.018	0.302	0.017	0.332	0.016	0.268	0.015	0.331	0.017	0.284	0.018	0.344	0.021	0.263	0.015
4	0.322	0.018	0.281	0.017	0.316	0.016	0.246	0.015	0.312	0.018	0.264	0.017	0.334	0.020	0.243	0.015
5	0.310	0.019	0.261	0.017	0.307	0.017	0.233	0.013	0.295	0.018	0.253	0.016	0.329	0.020	0.231	0.012
6	0.326	0.020	0.278	0.019	0.323	0.018	0.253	0.015	0.310	0.020	0.271	0.018	0.342	0.021	0.251	0.013
7	0.161	0.011	0.167	0.011	0.147	0.011	0.166	0.010	0.173	0.012	0.157	0.011	0.141	0.012	0.168	0.010
8	0.217	0.017	0.213	0.017	0.221	0.014	0.198	0.015	0.220	0.018	0.200	0.016	0.211	0.016	0.196	0.015
9	0.297	0.019	0.253	0.018	0.292	0.017	0.224	0.016	0.281	0.020	0.243	0.018	0.313	0.019	0.222	0.016
10	0.301	0.019	0.257	0.019	0.296	0.017	0.228	0.017	0.285	0.020	0.247	0.018	0.316	0.020	0.225	0.016
11	0.218	0.016	0.213	0.017	0.221	0.014	0.198	0.015	0.221	0.017	0.200	0.016	0.212	0.016	0.197	0.015
12	0.273	0.016	0.265	0.017	0.264	0.015	0.240	0.016	0.272	0.018	0.243	0.018	0.250	0.018	0.236	0.016
13	0.296	0.016	0.269	0.017	0.287	0.015	0.239	0.015	0.292	0.016	0.249	0.017	0.295	0.020	0.236	0.015
14	0.160	0.011	0.167	0.011	0.146	0.011	0.166	0.010	0.172	0.012	0.157	0.011	0.140	0.012	0.168	0.010
15	0.240	0.011	0.208	0.010	0.229	0.012	0.194	0.009	0.233	0.013	0.203	0.009	0.254	0.015	0.197	0.008
16	0.161	0.011	0.167	0.011	0.147	0.011	0.166	0.010	0.173	0.012	0.157	0.011	0.141	0.012	0.168	0.010
17	0.325	0.019	0.283	0.017	0.320	0.017	0.247	0.015	0.315	0.018	0.266	0.018	0.339	0.020	0.244	0.014
18	0.328	0.019	0.286	0.017	0.323	0.017	0.249	0.016	0.318	0.018	0.269	0.018	0.342	0.020	0.246	0.015
19	0.329	0.019	0.286	0.018	0.324	0.017	0.249	0.016	0.318	0.018	0.269	0.018	0.342	0.021	0.246	0.015
20	0.333	0.020	0.289	0.018	0.328	0.018	0.252	0.016	0.321	0.019	0.273	0.019	0.346	0.021	0.249	0.015
21	0.327	0.019	0.285	0.018	0.322	0.017	0.248	0.016	0.316	0.019	0.268	0.018	0.341	0.021	0.245	0.015
22	0.331	0.020	0.288	0.018	0.326	0.018	0.250	0.016	0.319	0.019	0.271	0.019	0.344	0.021	0.248	0.016
23	0.336	0.020	0.292	0.019	0.331	0.018	0.255	0.017	0.323	0.020	0.277	0.020	0.348	0.021	0.252	0.016
24	0.322	0.019	0.280	0.018	0.316	0.017	0.244	0.016	0.312	0.018	0.263	0.018	0.334	0.021	0.241	0.016
25	0.329	0.020	0.287	0.019	0.324	0.018	0.249	0.017	0.318	0.019	0.271	0.019	0.341	0.021	0.246	0.017
26	0.340	0.021	0.299	0.021	0.335	0.019	0.260	0.020	0.327	0.021	0.282	0.021	0.351	0.022	0.256	0.019
27	0.361	0.024	0.318	0.025	0.356	0.021	0.278	0.023	0.343	0.024	0.302	0.024	0.368	0.022	0.274	0.023
28	0.308	0.021	0.267	0.022	0.302	0.019	0.232	0.020	0.296	0.021	0.252	0.021	0.321	0.021	0.229	0.019
29	0.259	0.013	0.219	0.011	0.249	0.013	0.201	0.009	0.248	0.014	0.214	0.010	0.275	0.016	0.203	0.009
30	0.261	0.013	0.221	0.012	0.252	0.013	0.202	0.010	0.250	0.015	0.215	0.011	0.279	0.017	0.203	0.009
31	0.254	0.012	0.216	0.011	0.244	0.013	0.199	0.009	0.244	0.014	0.211	0.010	0.270	0.016	0.201	0.008
32	0.161	0.012	0.167	0.011	0.147	0.011	0.166	0.010	0.173	0.012	0.157	0.011	0.141	0.012	0.168	0.010
33	0.255	0.013	0.217	0.011	0.245	0.013	0.200	0.009	0.244	0.014	0.212	0.010	0.271	0.016	0.201	0.008
34	0.161	0.011	0.167	0.011	0.147	0.011	0.166	0.010	0.173	0.012	0.157	0.011	0.141	0.012	0.168	0.010
35	0.163	0.011	0.168	0.011	0.148	0.011	0.166	0.010	0.174	0.012	0.158	0.011	0.141	0.012	0.168	0.010
36	0.163	0.011	0.168	0.011	0.148	0.011	0.166	0.010	0.174	0.012	0.158	0.011	0.141	0.012	0.168	0.010
37	0.162	0.012	0.168	0.011	0.147	0.011	0.165	0.010	0.173	0.012	0.157	0.011	0.141	0.012	0.168	0.010
38	0.254	0.012	0.216	0.011	0.244	0.013	0.199	0.009	0.244	0.014	0.211	0.010	0.270	0.016	0.201	0.008
39	0.242	0.018	0.232	0.019	0.242	0.016	0.209	0.017	0.243	0.019	0.213	0.019	0.233	0.018	0.206	0.017
40	0.308	0.021	0.267	0.022	0.302	0.019	0.232	0.020	0.296	0.021	0.252	0.021	0.321	0.021	0.229	0.019
41	0.392	0.023	0.356	0.025	0.385	0.021	0.317	0.023	0.378	0.023	0.336	0.025	0.393	0.022	0.313	0.023
42	0.397	0.024	0.359	0.026	0.390	0.021	0.320	0.024	0.381	0.024	0.340	0.025	0.398	0.022	0.316	0.024
43	0.305	0.019	0.292	0.020	0.304	0.017	0.266	0.018	0.305	0.019	0.273	0.020	0.296	0.018	0.262	0.018
44	0.312	0.020	0.299	0.021	0.318	0.017	0.271	0.019	0.311	0.020	0.283	0.021	0.310	0.018	0.269	0.019
45	0.257	0.019	0.246	0.020	0.255	0.017	0.220	0.019	0.258	0.019	0.226	0.020	0.246	0.019	0.218	0.019
46	0.321	0.022	0.281	0.023	0.315	0.019	0.245	0.021	0.310	0.022	0.265	0.022	0.331	0.022	0.241	0.021

(continued)

Table 3. (*continued*)

	FFE								SWFE							
	Core		Default		Edge		GMM		Core		Default		Edge		GMM	
	Mean	Std	Mean	Std	Mean	Std	Mean	Std	Mean	Std	Mean	Std	Mean	Std	Mean	Std
47	0.316	0.022	0.276	0.022	0.310	0.019	0.240	0.021	0.305	0.021	0.260	0.022	0.327	0.021	0.237	0.020
48	0.259	0.019	0.248	0.020	0.256	0.017	0.222	0.019	0.260	0.019	0.228	0.020	0.247	0.019	0.219	0.019
49	0.255	0.019	0.245	0.020	0.254	0.017	0.219	0.019	0.256	0.019	0.226	0.020	0.245	0.019	0.217	0.019
50	0.257	0.019	0.246	0.020	0.255	0.017	0.220	0.019	0.258	0.019	0.226	0.020	0.246	0.019	0.218	0.019
51	0.391	0.019	0.363	0.019	0.380	0.018	0.328	0.018	0.381	0.019	0.340	0.020	0.383	0.021	0.323	0.018
52	0.389	0.018	0.360	0.019	0.378	0.017	0.326	0.018	0.379	0.019	0.338	0.020	0.381	0.021	0.322	0.017
53	0.387	0.018	0.359	0.019	0.376	0.017	0.325	0.017	0.378	0.019	0.336	0.020	0.379	0.021	0.320	0.017
54	0.384	0.018	0.356	0.018	0.373	0.017	0.322	0.017	0.375	0.018	0.333	0.019	0.376	0.021	0.318	0.017
55	0.383	0.017	0.355	0.018	0.373	0.016	0.327	0.016	0.374	0.018	0.335	0.018	0.387	0.019	0.324	0.016
56	0.381	0.018	0.351	0.019	0.372	0.017	0.319	0.017	0.371	0.019	0.329	0.019	0.381	0.020	0.315	0.017
57	0.374	0.019	0.343	0.020	0.364	0.018	0.308	0.018	0.364	0.019	0.321	0.020	0.371	0.021	0.304	0.018
58	0.367	0.020	0.333	0.020	0.359	0.018	0.298	0.019	0.357	0.020	0.313	0.021	0.367	0.021	0.294	0.019
59	0.343	0.019	0.312	0.021	0.333	0.018	0.277	0.019	0.335	0.019	0.290	0.021	0.342	0.021	0.273	0.019
60	0.320	0.020	0.284	0.021	0.314	0.018	0.248	0.019	0.311	0.020	0.266	0.021	0.328	0.021	0.244	0.019
61	0.258	0.019	0.247	0.020	0.256	0.017	0.221	0.019	0.259	0.019	0.227	0.020	0.247	0.019	0.218	0.019
62	0.262	0.019	0.250	0.020	0.258	0.017	0.224	0.019	0.262	0.019	0.229	0.020	0.248	0.019	0.220	0.019
63	0.319	0.022	0.280	0.023	0.314	0.019	0.243	0.021	0.309	0.022	0.264	0.022	0.330	0.022	0.240	0.021
64	0.305	0.019	0.270	0.020	0.299	0.018	0.235	0.018	0.298	0.019	0.252	0.020	0.315	0.021	0.232	0.018
65	0.265	0.018	0.255	0.018	0.260	0.016	0.230	0.017	0.266	0.018	0.234	0.019	0.249	0.018	0.226	0.017
66	0.266	0.018	0.256	0.018	0.261	0.016	0.231	0.017	0.267	0.018	0.235	0.019	0.250	0.018	0.227	0.017
67	0.266	0.018	0.256	0.019	0.261	0.016	0.230	0.017	0.267	0.018	0.234	0.019	0.250	0.018	0.226	0.018
68	0.265	0.018	0.255	0.019	0.260	0.016	0.229	0.017	0.266	0.018	0.233	0.019	0.249	0.018	0.225	0.018
69	0.268	0.018	0.257	0.018	0.261	0.016	0.231	0.017	0.268	0.018	0.235	0.019	0.250	0.018	0.227	0.017
70	0.320	0.020	0.284	0.021	0.314	0.018	0.248	0.019	0.311	0.020	0.266	0.021	0.328	0.021	0.244	0.019

Table 4. Load flow results for the transformer capacity for the first experiment.

	FFE								SWFE							
	Core		Default		Edge		GMM		Core		Default		Edge		GMM	
	Mean	Std	Mean	Std	Mean	Std	Mean	Std	Mean	Std	Mean	Std	Mean	Std	Mean	Std
0	0.381	0.019	0.348	0.019	0.371	0.017	0.3131	0.017	0.371	0.019	0.327	0.020	0.377	0.021	0.308	0.016

Table 5. Load flow results for the node voltage for the first experiment.

	FFE								SWFE							
	Core		Default		Edge		GMM		Core		Default		Edge		GMM	
	Mean	Std	Mean	Std	Mean	Std	Mean	Std	Mean	Std	Mean	Std	Mean	Std	Mean	Std
0	0.382	0.019	0.349	0.019	0.372	0.017	0.313	0.017	0.371	0.019	0.327	0.020	0.378	0.021	0.309	0.016
1	0.386	0.020	0.351	0.020	0.378	0.018	0.315	0.019	0.374	0.020	0.331	0.021	0.385	0.021	0.310	0.018
2	0.386	0.021	0.350	0.021	0.379	0.019	0.313	0.019	0.373	0.021	0.331	0.021	0.387	0.021	0.308	0.018
3	0.387	0.021	0.350	0.022	0.380	0.019	0.313	0.019	0.373	0.021	0.332	0.022	0.388	0.021	0.308	0.019
4	0.385	0.021	0.348	0.022	0.379	0.019	0.311	0.019	0.372	0.021	0.330	0.022	0.387	0.021	0.306	0.019
5	0.383	0.021	0.346	0.021	0.377	0.019	0.309	0.019	0.370	0.021	0.327	0.022	0.385	0.021	0.304	0.019
6	0.381	0.021	0.343	0.021	0.374	0.019	0.305	0.019	0.367	0.021	0.324	0.021	0.383	0.021	0.301	0.018
7	0.378	0.021	0.340	0.021	0.371	0.019	0.302	0.019	0.365	0.021	0.322	0.021	0.381	0.022	0.298	0.018

(*continued*)

Table 5. (*continued*)

	FFE								SWFE							
	Core		Default		Edge		GMM		Core		Default		Edge		GMM	
	Mean	Std	Mean	Std	Mean	Std	Mean	Std	Mean	Std	Mean	Std	Mean	Std	Mean	Std
8	0.377	0.021	0.338	0.021	0.370	0.019	0.301	0.019	0.363	0.021	0.320	0.021	0.380	0.022	0.296	0.018
9	0.377	0.021	0.338	0.021	0.370	0.019	0.300	0.019	0.363	0.021	0.320	0.021	0.380	0.022	0.296	0.018
10	0.388	0.021	0.353	0.021	0.381	0.019	0.316	0.019	0.376	0.021	0.334	0.022	0.389	0.021	0.312	0.019
11	0.382	0.021	0.346	0.021	0.375	0.019	0.309	0.019	0.369	0.020	0.327	0.021	0.383	0.021	0.305	0.018
12	0.389	0.022	0.352	0.022	0.383	0.019	0.315	0.020	0.376	0.021	0.334	0.022	0.391	0.021	0.311	0.020
13	0.386	0.022	0.349	0.022	0.380	0.019	0.312	0.020	0.373	0.021	0.331	0.022	0.388	0.021	0.307	0.019
14	0.383	0.021	0.346	0.021	0.376	0.019	0.309	0.019	0.370	0.021	0.327	0.021	0.385	0.021	0.304	0.018
15	0.381	0.021	0.343	0.021	0.374	0.019	0.306	0.019	0.367	0.020	0.325	0.021	0.383	0.021	0.301	0.018
16	0.376	0.021	0.337	0.021	0.369	0.019	0.300	0.019	0.362	0.020	0.319	0.021	0.379	0.022	0.295	0.018
17	0.374	0.021	0.336	0.021	0.368	0.019	0.298	0.019	0.361	0.020	0.318	0.021	0.378	0.021	0.293	0.018
18	0.377	0.021	0.338	0.021	0.371	0.019	0.300	0.019	0.363	0.021	0.320	0.022	0.381	0.022	0.295	0.018
19	0.360	0.018	0.323	0.018	0.352	0.017	0.287	0.016	0.350	0.018	0.304	0.019	0.362	0.021	0.282	0.016
20	0.352	0.019	0.314	0.018	0.344	0.017	0.277	0.016	0.341	0.018	0.295	0.018	0.357	0.021	0.272	0.016
21	0.348	0.019	0.309	0.018	0.341	0.017	0.272	0.016	0.337	0.019	0.290	0.018	0.354	0.021	0.267	0.016
22	0.345	0.019	0.305	0.018	0.338	0.017	0.268	0.016	0.334	0.019	0.287	0.019	0.352	0.021	0.263	0.016
23	0.343	0.019	0.302	0.018	0.336	0.017	0.265	0.016	0.332	0.019	0.284	0.019	0.351	0.021	0.261	0.016
24	0.341	0.019	0.301	0.018	0.335	0.017	0.263	0.016	0.330	0.019	0.283	0.019	0.350	0.021	0.259	0.016
25	0.341	0.019	0.300	0.018	0.335	0.017	0.263	0.016	0.330	0.019	0.282	0.019	0.350	0.021	0.259	0.016
26	0.340	0.019	0.299	0.018	0.334	0.017	0.262	0.016	0.329	0.019	0.281	0.019	0.349	0.021	0.258	0.016
27	0.340	0.019	0.299	0.018	0.334	0.017	0.262	0.016	0.329	0.019	0.281	0.019	0.349	0.021	0.258	0.016
28	0.340	0.019	0.299	0.018	0.334	0.017	0.262	0.016	0.329	0.019	0.281	0.019	0.349	0.021	0.258	0.016
29	0.341	0.019	0.300	0.018	0.335	0.017	0.263	0.017	0.330	0.019	0.282	0.019	0.350	0.021	0.259	0.016
30	0.341	0.019	0.301	0.018	0.335	0.017	0.263	0.017	0.330	0.019	0.283	0.019	0.350	0.021	0.259	0.016
31	0.355	0.018	0.318	0.018	0.347	0.017	0.282	0.016	0.345	0.018	0.299	0.018	0.358	0.021	0.277	0.015
32	0.349	0.018	0.311	0.018	0.342	0.017	0.274	0.016	0.339	0.018	0.292	0.018	0.355	0.021	0.270	0.016
33	0.346	0.019	0.306	0.018	0.339	0.017	0.270	0.016	0.335	0.018	0.288	0.018	0.353	0.021	0.265	0.015
34	0.345	0.019	0.305	0.018	0.339	0.017	0.268	0.016	0.334	0.019	0.287	0.019	0.353	0.021	0.263	0.016
35	0.341	0.019	0.301	0.018	0.335	0.017	0.264	0.016	0.331	0.019	0.283	0.018	0.350	0.021	0.260	0.016
36	0.340	0.019	0.299	0.018	0.333	0.017	0.262	0.016	0.329	0.019	0.281	0.018	0.349	0.021	0.258	0.016
37	0.343	0.019	0.302	0.018	0.336	0.018	0.264	0.017	0.331	0.019	0.284	0.019	0.352	0.021	0.260	0.016
38	0.339	0.019	0.298	0.018	0.332	0.017	0.261	0.016	0.328	0.019	0.280	0.018	0.348	0.021	0.257	0.016
39	0.339	0.019	0.298	0.018	0.333	0.017	0.261	0.016	0.328	0.019	0.281	0.019	0.349	0.021	0.257	0.016
40	0.339	0.019	0.298	0.018	0.333	0.017	0.261	0.016	0.328	0.019	0.280	0.019	0.348	0.021	0.257	0.016
41	0.343	0.019	0.302	0.018	0.336	0.018	0.264	0.017	0.331	0.019	0.284	0.019	0.351	0.021	0.260	0.016
42	0.341	0.019	0.301	0.018	0.335	0.017	0.263	0.017	0.330	0.019	0.283	0.019	0.350	0.021	0.259	0.016
43	0.387	0.019	0.354	0.020	0.377	0.018	0.317	0.018	0.376	0.019	0.332	0.021	0.382	0.021	0.313	0.018
44	0.389	0.020	0.356	0.021	0.380	0.018	0.319	0.019	0.377	0.020	0.335	0.021	0.385	0.021	0.314	0.018
45	0.388	0.020	0.354	0.021	0.379	0.019	0.317	0.019	0.376	0.020	0.334	0.021	0.385	0.021	0.313	0.019
46	0.389	0.020	0.354	0.021	0.380	0.019	0.317	0.019	0.376	0.020	0.334	0.022	0.386	0.021	0.313	0.019
47	0.389	0.021	0.354	0.021	0.380	0.019	0.317	0.019	0.376	0.021	0.334	0.022	0.386	0.021	0.313	0.019
48	0.385	0.019	0.351	0.020	0.375	0.018	0.315	0.018	0.374	0.019	0.330	0.021	0.381	0.021	0.310	0.018
49	0.394	0.021	0.359	0.022	0.385	0.019	0.322	0.020	0.381	0.021	0.339	0.022	0.391	0.021	0.318	0.019
50	0.386	0.020	0.352	0.021	0.376	0.018	0.314	0.019	0.374	0.020	0.331	0.021	0.382	0.021	0.310	0.018
51	0.389	0.021	0.354	0.021	0.380	0.019	0.316	0.020	0.376	0.021	0.333	0.022	0.386	0.021	0.312	0.019
52	0.389	0.021	0.354	0.021	0.380	0.019	0.316	0.019	0.376	0.021	0.333	0.022	0.386	0.021	0.311	0.019

(*continued*)

Table 5. (*continued*)

	FFE								SWFE							
	Core		Default		Edge		GMM		Core		Default		Edge		GMM	
	Mean	Std	Mean	Std	Mean	Std	Mean	Std	Mean	Std	Mean	Std	Mean	Std	Mean	Std
53	0.390	0.018	0.361	0.018	0.379	0.017	0.326	0.017	0.381	0.018	0.339	0.019	0.382	0.021	0.322	0.016
54	0.391	0.018	0.362	0.018	0.379	0.017	0.328	0.017	0.381	0.018	0.340	0.019	0.382	0.020	0.324	0.017
55	0.391	0.018	0.363	0.018	0.379	0.017	0.328	0.017	0.381	0.018	0.340	0.019	0.382	0.020	0.324	0.017
56	0.390	0.018	0.362	0.018	0.379	0.017	0.328	0.017	0.381	0.018	0.340	0.019	0.381	0.020	0.324	0.017
57	0.390	0.018	0.361	0.018	0.378	0.017	0.328	0.017	0.380	0.018	0.339	0.019	0.381	0.020	0.323	0.017
58	0.389	0.018	0.361	0.018	0.378	0.017	0.327	0.017	0.380	0.018	0.338	0.019	0.381	0.020	0.323	0.017
59	0.388	0.018	0.360	0.018	0.377	0.017	0.326	0.017	0.379	0.018	0.338	0.019	0.380	0.020	0.322	0.017
60	0.388	0.018	0.360	0.018	0.377	0.017	0.326	0.017	0.379	0.018	0.337	0.019	0.380	0.021	0.322	0.017
61	0.388	0.018	0.359	0.018	0.377	0.017	0.325	0.017	0.379	0.018	0.337	0.019	0.380	0.021	0.321	0.017
62	0.388	0.018	0.359	0.018	0.377	0.017	0.325	0.017	0.379	0.018	0.337	0.019	0.380	0.021	0.321	0.017
63	0.390	0.018	0.361	0.019	0.379	0.017	0.326	0.017	0.380	0.019	0.338	0.020	0.382	0.021	0.321	0.017
64	0.391	0.018	0.362	0.018	0.380	0.017	0.328	0.017	0.381	0.019	0.340	0.019	0.382	0.021	0.323	0.017
65	0.391	0.018	0.363	0.018	0.379	0.017	0.328	0.017	0.381	0.018	0.340	0.019	0.382	0.021	0.324	0.017
66	0.390	0.018	0.362	0.018	0.379	0.017	0.328	0.017	0.381	0.018	0.340	0.019	0.381	0.020	0.324	0.017
67	0.389	0.018	0.361	0.018	0.378	0.017	0.327	0.017	0.380	0.018	0.338	0.019	0.380	0.020	0.323	0.017
68	0.389	0.018	0.361	0.018	0.378	0.017	0.327	0.017	0.380	0.018	0.338	0.019	0.381	0.020	0.323	0.017
69	0.388	0.018	0.360	0.018	0.377	0.017	0.326	0.017	0.379	0.018	0.337	0.019	0.380	0.020	0.322	0.017
70	0.388	0.018	0.360	0.018	0.377	0.017	0.325	0.017	0.379	0.018	0.337	0.019	0.380	0.021	0.321	0.017

Appendix B   Detailed Results - Segmentwise Correlation

See Tables 6, 7 and 8

Table 6. Load flow results for the line capacity for the second experiment.

	FFE				SWFE			
	Core	Default	Edge	GMM	Core	Default	Edge	GMM
0	0.279	0.208	0.115	0.263	0.224	0.225	0.193	0.265
1	0.293	0.178	0.128	0.243	0.237	0.199	0.221	0.237
2	0.278	0.171	0.114	0.230	0.219	0.196	0.200	0.231
3	0.258	0.161	0.105	0.215	0.202	0.192	0.175	0.213
4	0.268	0.163	0.110	0.182	0.210	0.185	0.184	0.194
5	0.337	0.119	0.171	0.122	0.309	0.100	0.298	0.097
6	0.319	0.136	0.155	0.124	0.266	0.155	0.248	0.144
7	0.156	0.122	0.067	0.143	0.116	0.093	0.140	0.155
8	0.242	0.080	0.109	0.114	0.199	0.098	0.139	0.088
9	0.071	0.284	0.251	0.278	0.099	0.250	0.194	0.287
10	0.378	0.206	0.220	0.181	0.350	0.198	0.281	0.200

(*continued*)

Table 6. (*continued*)

	FFE				SWFE			
	Core	Default	Edge	GMM	Core	Default	Edge	GMM
11	0.214	0.076	0.123	0.102	0.183	0.089	0.159	0.081
12	0.143	0.101	0.083	0.053	0.113	0.073	0.111	0.082
13	0.084	0.175	0.221	0.240	0.111	0.209	0.182	0.228
14	0.170	0.136	0.060	0.202	0.123	0.122	0.148	0.124
15	0.136	0.212	0.190	0.169	0.122	0.176	0.107	0.185
16	0.156	0.122	0.067	0.143	0.116	0.093	0.140	0.155
17	0.305	0.197	0.132	0.290	0.251	0.247	0.239	0.289
18	0.310	0.191	0.135	0.284	0.258	0.242	0.247	0.281
19	0.310	0.184	0.135	0.281	0.257	0.238	0.246	0.274
20	0.313	0.172	0.140	0.272	0.265	0.225	0.254	0.262
21	0.316	0.157	0.139	0.254	0.269	0.210	0.257	0.247
22	0.318	0.147	0.140	0.242	0.272	0.194	0.258	0.237
23	0.309	0.146	0.137	0.235	0.257	0.193	0.246	0.227
24	0.295	0.138	0.117	0.223	0.240	0.183	0.221	0.218
25	0.294	0.141	0.115	0.217	0.240	0.177	0.218	0.196
26	0.304	0.121	0.114	0.197	0.248	0.156	0.222	0.185
27	0.255	0.164	0.140	0.227	0.211	0.177	0.154	0.222
28	0.331	0.149	0.113	0.092	0.282	0.137	0.235	0.081
29	0.279	0.131	0.115	0.122	0.233	0.159	0.182	0.093
30	0.344	0.184	0.170	0.156	0.299	0.155	0.269	0.161
31	0.226	0.105	0.119	0.125	0.202	0.110	0.139	0.082
32	0.104	0.085	0.133	0.127	0.068	0.108	0.058	0.151
33	0.343	0.218	0.169	0.154	0.312	0.173	0.266	0.182
34	0.170	0.126	0.055	0.117	0.118	0.074	0.135	0.111
35	0.136	0.091	0.044	0.123	0.089	0.053	0.121	0.087
36	0.158	0.103	0.050	0.062	0.109	0.084	0.141	0.089
37	0.139	0.067	0.047	0.046	0.097	0.043	0.122	0.067
38	0.443	0.327	0.284	0.268	0.406	0.326	0.406	0.322
39	0.103	0.190	0.220	0.264	0.094	0.201	0.121	0.257
40	0.331	0.149	0.113	0.092	0.282	0.137	0.235	0.081
41	0.351	0.166	0.165	0.192	0.301	0.148	0.298	0.185
42	0.338	0.140	0.162	0.158	0.301	0.129	0.287	0.149

(*continued*)

Table 6. (*continued*)

	FFE				SWFE			
	Core	Default	Edge	GMM	Core	Default	Edge	GMM
43	0.200	0.121	0.123	0.119	0.153	0.109	0.186	0.114
44	0.202	0.191	0.155	0.143	0.165	0.152	0.216	0.157
45	0.190	0.109	0.104	0.075	0.153	0.091	0.158	0.082
46	0.414	0.305	0.240	0.292	0.368	0.278	0.353	0.276
47	0.371	0.218	0.199	0.231	0.336	0.238	0.309	0.261
48	0.194	0.130	0.107	0.136	0.161	0.109	0.161	0.080
49	0.203	0.215	0.132	0.253	0.150	0.192	0.163	0.240
50	0.190	0.109	0.104	0.075	0.153	0.091	0.158	0.082
51	0.310	0.214	0.135	0.262	0.266	0.214	0.250	0.252
52	0.311	0.220	0.137	0.240	0.267	0.201	0.253	0.242
53	0.311	0.212	0.135	0.234	0.268	0.195	0.252	0.233
54	0.321	0.208	0.143	0.226	0.279	0.191	0.263	0.232
55	0.318	0.181	0.141	0.205	0.273	0.171	0.257	0.209
56	0.319	0.169	0.141	0.194	0.276	0.156	0.259	0.193
57	0.320	0.152	0.142	0.174	0.274	0.140	0.256	0.162
58	0.322	0.128	0.144	0.140	0.279	0.108	0.255	0.130
59	0.321	0.107	0.147	0.077	0.276	0.090	0.251	0.095
60	0.278	0.115	0.102	0.083	0.241	0.134	0.186	0.089
61	0.212	0.083	0.082	0.069	0.171	0.078	0.149	0.057
62	0.207	0.147	0.112	0.178	0.168	0.121	0.192	0.112
63	0.201	0.080	0.129	0.108	0.169	0.099	0.108	0.139
64	0.348	0.274	0.194	0.248	0.329	0.248	0.307	0.213
65	0.184	0.165	0.097	0.146	0.124	0.129	0.147	0.146
66	0.166	0.131	0.082	0.112	0.116	0.120	0.136	0.165
67	0.159	0.134	0.100	0.100	0.111	0.129	0.138	0.141
68	0.153	0.121	0.085	0.073	0.103	0.109	0.132	0.121
69	0.211	0.185	0.086	0.208	0.158	0.163	0.156	0.184
70	0.278	0.115	0.102	0.083	0.241	0.134	0.186	0.089

Table 7. Load flow results for the transformer capacity for the second experiment.

	FFE				SWFE			
	Core	Default	Edge	GMM	Core	Default	Edge	GMM
0	0.300	0.290	0.131	0.349	0.255	0.290	0.240	0.353

Table 8. Load flow results for the node voltage for the second experiment.

	FFE				SWFE			
	Core	Default	Edge	GMM	Core	Default	Edge	GMM
0	0.301058	0.290379	0.131642	0.349750	0.255887	0.290649	0.240938	0.353182
1	0.291083	0.264703	0.126491	0.310871	0.237296	0.272533	0.223898	0.310647
2	0.286720	0.237734	0.123947	0.304424	0.230838	0.264110	0.218171	0.302667
3	0.287901	0.223846	0.125376	0.298438	0.231287	0.254155	0.218321	0.295549
4	0.286785	0.213548	0.124261	0.293641	0.229490	0.246680	0.216851	0.290151
5	0.285535	0.205359	0.123280	0.290673	0.228203	0.240959	0.215380	0.287281
6	0.285237	0.196890	0.122655	0.289210	0.227231	0.238491	0.214485	0.285545
7	0.287154	0.193539	0.123415	0.286068	0.229588	0.234965	0.217831	0.284139
8	0.287631	0.193449	0.123848	0.284377	0.229453	0.233632	0.218293	0.284199
9	0.289506	0.191663	0.125213	0.282610	0.233069	0.231982	0.221908	0.282846
10	0.292447	0.250082	0.126385	0.310119	0.237374	0.270486	0.223428	0.307875
11	0.278509	0.220925	0.116254	0.303615	0.222137	0.267069	0.201712	0.304386
12	0.290793	0.217580	0.127067	0.296521	0.234363	0.253929	0.221524	0.291909
13	0.288334	0.210756	0.125098	0.292485	0.230981	0.243126	0.218184	0.288736
14	0.282949	0.200911	0.121838	0.288097	0.225760	0.239022	0.213178	0.283646
15	0.281528	0.198210	0.119073	0.290494	0.222819	0.238297	0.206819	0.286749
16	0.289506	0.193300	0.125106	0.284178	0.235748	0.232108	0.223656	0.281425
17	0.286025	0.196423	0.121675	0.283719	0.228047	0.233642	0.216091	0.284464
18	0.291091	0.191127	0.126250	0.280729	0.236046	0.229795	0.225283	0.281087
19	0.300993	0.238076	0.130662	0.320584	0.253594	0.276004	0.239244	0.320131
20	0.303195	0.216226	0.131585	0.316826	0.255091	0.270403	0.241272	0.315642
21	0.303187	0.213394	0.131051	0.313244	0.254655	0.267121	0.241945	0.312098
22	0.305376	0.208984	0.133154	0.309708	0.256655	0.262292	0.245870	0.306072
23	0.305903	0.203899	0.134363	0.305905	0.258179	0.258141	0.247885	0.300973
24	0.307812	0.200924	0.135491	0.302274	0.259595	0.254468	0.249520	0.296213
25	0.307977	0.198683	0.135888	0.299803	0.260397	0.252404	0.250230	0.293347
26	0.307195	0.196863	0.134785	0.299096	0.258444	0.250356	0.248256	0.291406
27	0.306754	0.195606	0.133099	0.297895	0.258036	0.247978	0.247330	0.290054
28	0.306577	0.194281	0.132436	0.297489	0.255626	0.247368	0.245091	0.289022
29	0.306173	0.193645	0.131009	0.297094	0.252826	0.246289	0.241922	0.288715
30	0.306505	0.192606	0.130817	0.296373	0.252757	0.245608	0.241338	0.288110
31	0.298109	0.225769	0.129637	0.316950	0.248889	0.271795	0.234925	0.319965
32	0.303937	0.214745	0.131393	0.314450	0.256434	0.268476	0.242499	0.313260
33	0.302692	0.213512	0.129674	0.312919	0.252186	0.265704	0.239373	0.312805
34	0.305234	0.208798	0.133702	0.309213	0.255036	0.261518	0.243761	0.307008
35	0.305778	0.202679	0.134200	0.305028	0.257412	0.257547	0.246742	0.299863

(continued)

Table 8. (*continued*)

	FFE				SWFE			
	Core	Default	Edge	GMM	Core	Default	Edge	GMM
36	0.308343	0.200287	0.136194	0.300773	0.261803	0.253132	0.250721	0.294879
37	0.310149	0.197262	0.137138	0.298157	0.262397	0.251237	0.252518	0.291663
38	0.306912	0.195793	0.133976	0.298311	0.257738	0.249386	0.247903	0.290573
39	0.305965	0.195323	0.133253	0.297472	0.256760	0.247230	0.246871	0.289364
40	0.307550	0.192475	0.133253	0.296468	0.257204	0.245798	0.246375	0.287838
41	0.304837	0.193650	0.130148	0.298438	0.251124	0.246920	0.239728	0.289647
42	0.306230	0.190558	0.130166	0.295156	0.252819	0.244149	0.241028	0.287107
43	0.309171	0.279089	0.136708	0.334430	0.269729	0.276637	0.253133	0.330415
44	0.311842	0.260455	0.138313	0.306930	0.275895	0.251904	0.258719	0.307511
45	0.313113	0.247051	0.138746	0.297660	0.275643	0.242640	0.258481	0.289387
46	0.314470	0.237019	0.139833	0.292154	0.279282	0.236243	0.261962	0.284151
47	0.312488	0.223281	0.139485	0.289888	0.277186	0.233443	0.259702	0.280810
48	0.312288	0.264351	0.138668	0.320465	0.274862	0.253981	0.257379	0.312772
49	0.315157	0.247783	0.140139	0.297322	0.281987	0.243660	0.266005	0.294101
50	0.310569	0.235212	0.137688	0.294523	0.270004	0.237872	0.253024	0.284973
51	0.318734	0.229955	0.142100	0.283809	0.283714	0.228011	0.268192	0.276684
52	0.311210	0.203422	0.139192	0.286273	0.274347	0.229660	0.257027	0.277112
53	0.304731	0.271182	0.133673	0.327885	0.260702	0.265039	0.244959	0.331190
54	0.306990	0.256711	0.135180	0.304464	0.263359	0.248537	0.247118	0.305284
55	0.307153	0.251082	0.135450	0.292297	0.263581	0.239890	0.247688	0.291185
56	0.308951	0.247653	0.135911	0.283997	0.266360	0.234011	0.251011	0.282129
57	0.309943	0.243759	0.136282	0.275671	0.267553	0.229484	0.252020	0.276849
58	0.311614	0.238027	0.136728	0.269787	0.269336	0.225301	0.253487	0.270190
59	0.313004	0.233165	0.136970	0.267218	0.270030	0.221680	0.253612	0.262524
60	0.313388	0.230584	0.137595	0.265366	0.270130	0.219040	0.252748	0.256945
61	0.314143	0.227446	0.137766	0.262909	0.270341	0.217142	0.252382	0.255830
62	0.313121	0.222560	0.136970	0.262188	0.269050	0.216472	0.251610	0.255644
63	0.303768	0.256099	0.132978	0.317528	0.258366	0.257373	0.241700	0.319244
64	0.307068	0.251253	0.135266	0.302271	0.263277	0.244213	0.248269	0.302498
65	0.305249	0.248297	0.134093	0.282822	0.261217	0.239718	0.244937	0.286997
66	0.309466	0.245068	0.136487	0.282950	0.267472	0.232892	0.252056	0.279926
67	0.310320	0.242378	0.136775	0.273262	0.267578	0.226641	0.252055	0.275848
68	0.311627	0.236518	0.137076	0.268346	0.269223	0.223601	0.253928	0.269909
69	0.313123	0.230747	0.136863	0.267109	0.269711	0.219962	0.253376	0.261546
70	0.313297	0.229099	0.137317	0.264667	0.270105	0.217712	0.253111	0.255741

References

1. Henze, J., Kneiske, T., Braun, M., Sick, B.: Identifying representative load time series for load flow calculations. In: Woon, W.L., Aung, Z., Kramer, O., Madnick, S. (eds.) DARE 2017. LNCS (LNAI), vol. 10691, pp. 83–93. Springer, Cham (2017). https://doi.org/10.1007/978-3-319-71643-5_8
2. Green, R., Staffell, I., Vasilakos, N.: Divide and Conquer? k-means clustering of demand data allows rapid and accurate simulations of the British electricity system. IEEE Trans. Eng. Manag. **61**(2), 251–260 (2014)
3. Poncelet, K., Hoschle, H., Delarue, E., Virag, A., D'haeseleer, W.: Selecting representative days for capturing the implications of integrating intermittent renewables in generation expansion planning problems. IEEE Trans. Power Syst. **32**(3), 1936–1948 (2017)
4. Räsänen, T., Kolehmainen, M.: Feature-based clustering for electricity use time series data. In: Kolehmainen, M., Toivanen, P., Beliczynski, B. (eds.) ICANNGA 2009. LNCS, vol. 5495, pp. 401–412. Springer, Heidelberg (2009). https://doi.org/10.1007/978-3-642-04921-7_41
5. Li, X., Bowers, C.P., Schnier, T.: Classification of energy consumption in buildings with outlier detection. IEEE Trans. Ind. Electron. **57**(11), 3639–3644 (2010)
6. Merrick, J.H.: On representation of temporal variability in electricity capacity planning models. Energy Econ. **59**, 261–274 (2016)
7. Hajian, M., Rosehart, W.D., Zareipour, H.: Probabilistic power flow by Monte Carlo simulation with Latin supercube sampling. IEEE Trans. Power Syst. **28**, 1550–1559 (2013)
8. Kabir, M., Mishra, Y., Bansal, R.: Probabilistic load flow for distribution systems with uncertain PV generation. Appl. Energy **163**, 343–351 (2016)
9. Mitra, J., Singh, C.: Incorporating the DC load flow model in the decomposition-simulation method of multi-area reliability evaluation. IEEE Trans. Power Syst. **11**(3), 1245–1254 (1996)
10. Wang, Y., Chen, Q., Kang, C., Zhang, M., Wang, K., Zhao, Y.: Load profiling and its application to demand response: a review. Tsinghua Sci. Technol. **20**(2), 117–129 (2015)
11. Fischer, D., Härtl, A., Wille-Haussmann, B.: Model for electric load profiles with high time resolution for German households. Energy Build. **92**, 170–179 (2015)
12. McInnes, L., Healy, J., Astels, S.: hdbscan: Hierarchical density based clustering. J. Open Source Softw. Open J **2**(11) (2017)
13. Kerber, G.: Aufnahmefähigkeit von Niederspannungsverteilnetzen für die Einspeisung aus Photovoltaikkleinanlagen. Dissertation, Technische Universität München (2011)
14. Lindner, M., Aigner, C., Witzmann, R., Wirtz, F., Berber, I., Gödde, M., Frings, R.: Aktuelle musternetze zur untersuchung von spannungsproblemen in der niederspannung. In: 14. Symposium Energieinnovation (2016)
15. Thurner, L., Scheidler, A., Schafer, F. (et.al): Pandapower - an open source python tool for convenient modeling, analysis and optimization of electric power systems. IEEE Trans. Power Syst. **33**(6), 6510–6521 (2018)
16. Kolmogorov, A.: On the empirical determination of a distribution function. In: Kotz, S., Johnson, N.L. (eds.) Breakthroughs in Statistics. SSS, pp. 106–113. Springer, New York (1992). https://doi.org/10.1007/978-1-4612-4380-9_10

Probabilistic Graphs for Sensor Data-Driven Modelling of Power Systems at Scale

Francesco Fusco[(✉)]

IBM Research - Ireland, Dublin, Ireland
francfus@ie.ibm.com

Abstract. The growing complexity of the power grid, driven by increasing share of distributed energy resources and by massive deployment of intelligent internet-connected devices, requires new modelling tools for planning and operation. Physics-based state estimation models currently used for data filtering, prediction and anomaly detection are hard to maintain and adapt to the ever-changing complex dynamics of the power system. A data-driven approach based on probabilistic graphs is proposed, where custom non-linear, localised models of the joint density of subset of system variables can be combined to model arbitrarily large and complex systems. The graphical model allows to naturally embed domain knowledge in the form of variables dependency structure or local quantitative relationships. A specific instance where neural-network models are used to represent the local joint densities is proposed, although the methodology generalises to other model classes. Accuracy and scalability are evaluated on a large-scale data set representative of the European transmission grid.

Keywords: Probabilistic graphical models · Factor graphs
Data imputation · Anomaly detection · Power system state estimation

1 Introduction

The increasingly large share of renewable energy generation being injected into the electrical grid, along with the growing diversity and complexity in the mix of distributed resources necessary for balancing supply and demand (e.g. home energy management systems, smart thermostats, electric storage and vehicles) call for novel power system modelling tools. Operational and planning decisions require visibility into a wider range of variables as opposed to the limited view of net power flows and injections available through traditional tools based on first-principle equations, such as power-flow and state-estimation models [3]. Furthermore, expanding and maintaining such physics-based tools, which involves identifying and updating the right sets of fundamental equations and physical parameters, becomes increasingly harder given the higher degree of complexity of the system dynamics and its fast-changing nature.

© Springer Nature Switzerland AG 2018
W. L. Woon et al. (Eds.): DARE 2018, LNAI 11325, pp. 49–62, 2018.
https://doi.org/10.1007/978-3-030-04303-2_4

The unprecedented availability of data, resulting from increasing numbers of Internet-of-Things (IOT) sensor devices [5], makes data-driven approaches to power system modelling practical. Besides its inherent ability to adapt and grow along with the available data, a data-driven model should serve the key functionalities that are typical of power system state estimation tools [3]:

- run model predictions based on assumptions about a subset of variables (short-term grid predictions, what-if scenarios for planning and optimization);
- filter observations to reduce sensor noise and reconstruct missing data;
- detect and localise system or data anomalies.

Numerous classes of machine-learning approaches can be used to model the non-linear power system dynamics and handle data imputation, sensor noise and outliers. Variational autoencoders provide an efficient approach to denoising [15], although they do not deal naturally with missing data. Non-linear extensions, based on feedforward neural networks, of principal component analysis (NLPCA) [20] and factor analysis (NLFA) [11] are well suitable to solve the data imputation problem. Non-linear extentions of PCA based on kernel methods [18,19] or Gaussian processes [17] can also be used to handle missing data. One limitation of such models, however, is that they cannot naturally account for the factorization and sparsity induced by power systems topology. Even where efficient computational frameworks exist to handle large number of variables, for example in the case of neural networks, the consequent exponential increase in parameters makes their demands in terms of training data not practical in very dynamic and changing sensor data-driven systems.

More recently, computational methods based on probabilistic graphical models and belief propagation were studied to offer a naturally distributed solution to the traditional power systems state-estimation problem [6,7,12]. A belief-propagation algorithm supporting a combination of computational nodes representing state-estimation sub-problems, as well as other custom non-linear relations, was proposed in [10]. All research on probabilistic graphs applied to power systems, however, focused on representing power-flow equations with a graphical model and on the inference task. While offering improved flexibility, such models still rely on given sets of fundamental equations and knowledge of their physical parameters. Keeping track of the updates in grid topology and parameters is a challenging, time-consuming and at times impossible, task. None of the mentioned research deals with the problem of learning the model relationships fully from the data. In this study, a learning algorithm is derived based on the assumption that the non-linear joint densities between subsets of variables are modelled as Gaussian latent-variable models, using a neural-network NLPCA model [20]. It is demonstrated how the NLPCA factor nodes naturally handle missing data, thus being well suited for the mentioned power system modelling tasks, which can mostly be reduced to data imputation problems. It is also shown how very simple heuristics based on the power system topology can be used to initialize the graphical model structure such to obtain a more efficient model (in terms of number of parameters) than a centralised NLPCA with a similar modelling accuracy. It is also shown how, in the presence limited training data, the accuracy of the proposed graphical model does not degrade with increasing dimensionality of the problem, in contrast to the centralised approach.

The model is detailed in Sect. 2. Results on a publicly-available data set representative of the European transmission grid, are reported in Sect. 3. Conclusive remarks and scope for further work are outlined in Sect. 4.

2   Methodology

Graphical models provide a natural way to represent structural relationships between random variables, to embed domain knowledge and to deal with missing data. The proposed method is derived using the factor graphs representation, which provides a way to link, in the form of conditional and joint probability distributions, potentially very different models together in a principled fashion that respects the rules of probability theory [9].

After introducing the modelling approach and assumptions, in Sect. 2.1, the neural-network NLPCA approach used to model the non-linear joint densities is detailed in Sect. 2.2. The inference and learning algorithms are then described in Sects. 2.3 and 2.4.

2.1   Model

Given a set of variables, $\mathcal{Z} = \{z_1, z_2, \ldots\}$, graphical models are a mathematical tool which can be used to express factorizations of the form $\Phi(z_1, z_2, \ldots) = \prod_j \phi_j(\mathcal{Z}_j)$, where $\phi_j(\mathcal{Z}_j)$ is a probability density defined on a subset of the variables $\mathcal{Z}_j \in \mathcal{Z}$. In a factor graph parametrization, a *variable node* is defined for each variable z_i and a *factor node* represents the densities ϕ_j. A graph edge between a factor node and a variable node exists for each $z_i \in \mathcal{Z}_j$. Factor nodes can represent both conditional densities (directed edges) and joint densities (undirected edges), thus offering a very general modelling framework.

For the specific purposes outlined in Sect. 1, the set of variables $\mathcal{Z} = \{\mathcal{X}, \mathcal{Y}\}$ is a combination of N random state variables $\mathcal{X} = \{x_1, x_2, \ldots, x_N\}$ and M sources of noisy observations $\mathcal{Y} = \{y_1, y_2, \ldots, y_M\}$. The joint probability distribution is factorized as:

$$p(\mathcal{X}, \mathcal{Y}) = \prod_{m=1}^{M} p(y_m | \mathcal{X}_m) \prod_{k=1}^{K} p(\mathcal{X}_k), \qquad (1)$$

where $p(y_m|\mathcal{X}_m)$ denotes a conditional probability density, $p(\mathcal{X}_k)$ is a joint probability density and $\mathcal{X}_i \in \mathcal{X}$. It is assumed that both types of relationships can be modelled with Gaussian densities as follows:

$$p(\mathcal{X}, \mathcal{Y}) \propto \prod_{m=1}^{M} e^{-\frac{1}{2}[y_m - f_m(\mathcal{X}_m)]^\top R_m [y_m - f_m(\mathcal{X}_m)]} \prod_{k=1}^{K} e^{-\frac{1}{2} g_k(\mathcal{X}_k)^\top S_k g_k(\mathcal{X}_k)}, \qquad (2)$$

where $f_m(\cdot)$, $g_k(\cdot)$ are generally non-linear functions and R_i, S_k are covariance matrices. The Gaussian assumption allows for efficient inference and learning algorithms to be designed, as shown in Sects. 2.3 and 2.4, and it is reasonably accurate when applied to sensor data processing in power systems [2].

In (2), the conditional densities express explicit relations between the data and the state variables. They can be used to embed domain knowledge into the model such that the state variable has a desired physical meaning. Typically, in power systems, the state variables are defined as the set of voltage magnitudes and angles, and all data are related to them through power flow equations parametrized on the topology and impedances of the grid. In a fully data-driven model, however, any other desired representation is possible and, for example, state variables could include, for example, amount of renewable generation, demand- response capacity at the nodes of the grid, as long as an explicit relation to the data exists, as it will be demonstrated in Sect. 3.

The joint probabilities in (2) represent structural relationships between subset of state variables within the model. Domain knowledge and heuristics based on power system topology can give indications on the structure of the joint densities. However, the functional form of the relation, $g(\cdot)$, alon with the covariance S_k, is not known in general and should be learned from the data. It is reasonable to assume that the joint density of a set of state variables can be fully specified by a lower-dimensional latent variable, that is $p(\mathcal{X}_k) = p(\mathcal{X}_k|z_k)$. In Sect. 2.2, an approach to model such densities using NLPCA, originally proposed in [20], is described. The methodology, however, is quite general and any other class of latent-variable models, as reviewed in Sect. 1, could be used at no loss of generality for the inference and learning algorithms derived in Sects. 2.3 and 2.4.

2.2   Joint Factor Node Based on NLPCA

The NLPCA model considered here is a modification of the classical autoencoder, where only the decoder neural network is learnt from the data, while the encoding function is based on actual model inversion through gradient descent [20], thus providing for a more natural way for handling missing data.

A feed-forward neural network provides a non-linear mapping of the type:

$$\mathcal{X}_k = h_k(z_k, \vartheta_k) + \zeta_k, \tag{3}$$

where it is assumed that $\zeta_k \sim \mathcal{N}(0, R_k)$ is a zero-mean Gaussian error term with covariance matrix R_k and ϑ_k are the network parameters (weights). Note that the input z_k is also unknown and an extension to back-propagation is needed for learning both ϑ_k and z_k at the same time [20]. By including an additional input layer, such that the input to the network is an identity matrix and the weights of the input layer represent the values of z_k, the NLPCA model can be conveniently trained with conventional back-propagation [20].

Inference requires to run back-propagation, with ϑ_k fixed, in order to estimate the latent variable z_k for a given value of $\mathcal{X}_k$. Such learning, however, is a relatively low-dimensional optimization with respect to the learning stage, and it typically runs very efficiently [20]. Missing data are dealt with quite naturally since it suffices to remove the missing components of $\mathcal{X}_k$ from the gradient calculations.

With respect to the joint density $p(\mathcal{X}_k) \propto e^{-\frac{1}{2}g_k(\mathcal{X}_k)^\top S_k g_k(\mathcal{X}_k)}$ in (2), a factor node based on the NLPCA model is ultimately defined by:

$$g_k(\mathcal{X}_k) = h_k(h_k^{-1}(\mathcal{X}_k)), \tag{4}$$

$$S_k = (\tilde{H}_k^\top R_k^{-1} \tilde{H}_k)^{-1}. \tag{5}$$

As mentioned, the model inversion in (4), $h^{-1}(\cdot)$, has no explicit form but it represents a gradient-descent procedure. The expression for the covariance matrix in (5) makes use of a typical approximation from non-linear Gaussian filtering [3], where $\tilde{H}_k$ is the Jacobian of the encoding function computed at the solution for the latent variable, namely $\nabla h(z_k)$. The notation $\tilde{H}_k$ is used to highlight the fact that, in the case of partially available data, $\tilde{H}_k$ is only a subset of the full gradient H_k, composed of the rows corresponding to the subset of available data.

The adopted NLPCA model is not a generative model in the sense that no model for $p(z_k)$ is provided, but only an approximation of $p(z_k|\mathcal{X}_k)$. As a consequence, inference can only be solved in the case of partially missing data. While this is generally a limitation, generative models are typically more computationally intensive and, more importantly, there are no practical scenarios for the case of fully missing data. The proposed treatment of probabilistic graphs is, however, not restricted to the particular choice for the latent variable model and generative models, for example based on NLFA [11], can be used instead.

2.3 Inference

Running inference on graphical models reduces to the computation of localised messages between factor and variable nodes, along each edge of the graph defined by the factorization of the density function in (1) [4]. In the particular case of Gaussian joint densities, the process of belief propagation involves Gaussian messages and simplifies to the sum-product algorithm, that is a set of summations and multiplications over the parameters of the Gaussian distributions [16].

It is convenient to express the factor densities in the canonical form, $p(x) \propto e^{-\frac{1}{2}x^\top Jx + x^\top h}$. Linearization based on a first-order Taylor expansion yields:

$$J_m = F_m^\top R_m^{-1} F_m \tag{6}$$

$$h_m = F_m^\top R_m^{-1}(y_m - f_m(\overline{\mathcal{X}}_m)) \tag{7}$$

for the conditional densities and

$$J_k = G_k^\top S_k^{-1} G_k \tag{8}$$

$$h_k = G_k^\top S_k^{-1}(\mathcal{X}_k - g_k(\overline{\mathcal{X}}_k)) \tag{9}$$

for the joint densities. In (6) to (9), F_m and G_k are the Jacobian matrices, respectively, of the functions $f_m(\mathcal{X}_k)$ and $g_k(\mathcal{X}_k)$ computed with respect to a value of the state variable $\overline{\mathcal{X}}_k$.

As derived in [10,16], messages from variable x_i to factor f_j are computed as:

$$h_{x_i \to f_j} = \sum_{k \in \mathcal{K}_i \setminus j} h_{f_k \to x_i} \tag{10}$$

$$J_{x_i \to f_j} = \sum_{k \in \mathcal{K}_i \setminus j} J_{f_k \to x_i}, \tag{11}$$

where $\mathcal{K}_i$ is the set of factors f_i connected to the variable x_i. Messages from factor j to variable i are calculated based on:

$$h_{f_j \to x_i} = h_j - \sum_{k \in \mathcal{K}_j \setminus i} J_j^{jk} (J_{x_k \to f_j} + J_j^{kk})^{-1} (h_{x_k \to f_j} + h_j^k) \tag{12}$$

$$J_{f_j \to x_i} = J_j - \sum_{k \in \mathcal{K}_j \setminus i} J_j^{jk} (J_{x_k \to f_j} + J_j^{kk})^{-1} J_j^{kj}, \tag{13}$$

where J_j^{jk} is the block of the J_j matrix with rows corresponding to the variable x_j and columns corresponding to variable x_k. Similarly, h_j^k denotes the block of the h_j vector corresponding to the variable x_j.

Under the assumption of tree-structured graph, all messages (10) to (13) can be computed by starting from the leaf factor nodes and iteratively updating all messages where the required input messages have been processed. If the graph is not a tree, a loopy version of the proposed belief propagation algorithm can be derived to iteratively converge to a solution [7]. Once messages from all incoming factors are available, variable estimates can be updated with the following iterative scheme:

$$x_i^{t+1} = x_i^t + \delta x_i \tag{14}$$

$$\delta x_i = \left(\sum_{k \in \mathcal{K}_i} J_{f_k \to x_i} \right)^{-1} \left(\sum_{k \in \mathcal{K}_i} h_{f_k \to x_i} \right), \tag{15}$$

with the covariance matrix given by:

$$S_{x_i}^{t+1} = \left(\sum_{k \in \mathcal{K}_i} J_{f_k \to x_i} \right)^{-1}. \tag{16}$$

It is interesting to note that, for a tree-structured graph with n variable nodes of dimension d, the computational complexity of each iteration (without accounting for the back-propagation in a NLPCA joint factor node) is $o(nd^3)$, as opposed to $o(n^3 d^3)$ that would be required by working on the global density directly. Being able to represent the factorization of the problem, which is quite a natural property of physical interconnected, distributed systems such as electrical grids, probabilistic graphs have excellent scalability properties.

2.4   Learning

Based on the model definition in Sect. 2.1, it is assumed that sensor data are available for the variables in $\mathcal{Y}$. Given a data set $\mathcal{D}$ of N observations $\{y_1^{1:N}, \ldots y_m^{1:N}\}$, the graphical model in (2) is trained by solving the following maximum-likelihood problem:

$$\underset{\vartheta, \gamma}{\operatorname{argmax}} \, L(\vartheta, \gamma | \mathcal{D}) = \prod_{n=1}^{N} \prod_{m=1}^{M} p(y_m^n | \mathcal{X}_m^n, \vartheta_m) \prod_{k=1}^{K} p(\mathcal{X}_k^n | z_k^n, \gamma_k), \qquad (17)$$

where ϑ are unknown parameters of the conditional densities and γ are parameters of the joint densities, represented as conditional, latent variable models, for example using the NLPCA approach proposed in Sect. 2.2.

The state variables in $\mathcal{X}$ are latent variables and are never directly observed. Expectation maximization is therefore used to solve (17) by iteratively running an inference to solve for all the latent (or missing) variables and then maximising the likelihood. The Gaussian factorization results in localised likelihood functions so that, given an estimate for the latent variables, the parameters ϑ_m and γ_k can be solved for by running independent maximum likelihood on each individual factor density [16]. Specifically in the case of the NLPCA models, a modified gradient descent minimising the squared error is applied, as described in Sect. 2.2. Similarly, an independent estimator for each parametric conditional density can be designed. In the present study, the conditional densities are assumed to be known as simple identity mapping from the observations to the latent variables, such that $\mathcal{X}$ assumes a well-defined physical meaning. Learning of the conditional densities is therefore not considered here.

3   Results

3.1   Data Preparation and Model Design

A publicly available large-scale dataset, created for modelling electricity demand and renewable generation in the European transmission grid, was used [13]. The data set includes hourly time series data of energy demand, solar generation and wind generation at nearly 1500 points across Europe from the year 2012 through to 2014. The data set was augmented by running a power-flow simulation to generate time series of voltage and reactive power injections, using the 1354-bus electrical model of the European transmission grid made available in [14] and the Matpower software [21]. The bus active power injections where generated by combining loads, wind and solar generation signals at random locations of the data in [13], while the reactive power injections were based on a 0.9 power factor. Sensor noise was simulated as zero-mean, Gaussian random error with standard deviation of $1e^{-3}$ for power measurements and of $1e^{-5}$ for voltage measurements. As a result, a set of 8124 hourly time series data of active/reactive power, voltage magnitude, energy demand, solar and wind generation at 1354 locations was

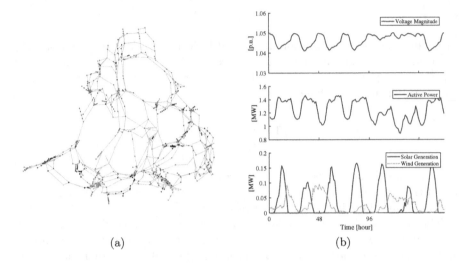

      (a)   (b)

Fig. 1. (a) Connectivity of the 1354-bus power grid. The color-coding represents the 273 graph partitions obtained using the spectral method based on Fiedler eigenvectors, as described in Sect. 3.1. (b) Sample data of voltage magnitude, active power, solar and wind generation at a grid location. (Color figure online)

available, for 3 years. Figure 1 shows the grid topology and sample time-series data at one grid location.

The factorization of the proposed graphical model was defined from the grid connectivity based on simple heuristics. The network graph is partitioned using the laplacian of the connectivity matrix and the associated Fiedler eigenvectors [8], by iteratively bi-sectioning according to the eigenvector corresponding to the second smallest eigenvalue of the Laplacian matrix. The color-coding in Fig. 1(a) indicates the partitioning obtained after 9 iterations, which resulted in 273 sections with a number of nodes ranging from 1 to 28. In relation to the model described in Sect. 2.1, a variable node representing the set of electrical quantities (voltage, active/reactive power, demand, solar/wind generation) at the nodes of each partition was defined. A conditional factor for each metered time series was defined as a simple identity mapping, with a covariance matrix based on the assumed sensor noise. A NLPCA-based joint factor node was introduced for each pair of variable nodes of directly connected graph partitions. The NLPCA model of each joint factor node was based on a 3-layered, fully-connected, neural network with a linear output layer and a sigmoid activation in the hidden layer. The dimensionality of the latent variable was set to half the dimensionality of the output. Loops in the graph were avoided by only connecting variables to at most two joint factors. The final topology of the proposed graphical model is shown in Fig. 2. Note that many other possible factorizations and corresponding topological structures could be defined, based on alternative heuristics or data-driven methods. Such investigation is not in scope of the current study and the proposed structure, named as Graph-NLPCA, compares favourably with a centralised model using NLPCA, as discussed in the Sects. 3.2, 3.3 and 3.4.

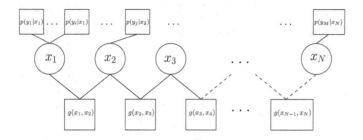

Fig. 2. Topology of the proposed graphical model.

As discussed in Sect. 2.4, model training is based on a few iterations of inference on the factor graph and learning of the individual NLPCA models. Given the flat structure of the model, it was found that less than 5 iterations are sufficient to converge to an accurate model. Back-propagation training and inference on the NLPCA models was performed using the TensorFlow computational library, using mean square error objective function and stochastic gradient descent [1]. Data for the month of July 2014 were used for training, and validation was done on the month of August 2014. This is to reflect typical applications where the power system dynamics change over time, due to topological changes (grid expansion, reinforcements) and power flow variations (new customer connections, new renewable generation installations), and it is important that the models are updated on a regular basis only on a short history of the data. Note that while electrical load and generation time series typically express strong seasonality, that is only expected to play a limited role, given the spatial nature of the model. In particular, the proposed model is aimed at capturing the relationship between grid quantities at different locations, rather than the dynamics of a given electrical quantity at different points in time.

3.2 Model Accuracy on Missing Data

As outlined in Sect. 1, the primary objective of a data-driven power system model is to being able to handle missing data, such to solve both as a prediction model and sensor data filtering functions. A range of experiments was designed to evaluate the estimation error of the proposed Graph-NLPCA under different proportions of missing data. As a baseline, a centralised approach where all the variables are included in a single NLPCA model is considered. Only part of the complete transmission grid model in the available data is considered, made of 10 grid locations for a total of 375 variables. For this particular dimensionality of the problem, the number of parameters of a NLPCA model, 211500, is comparable with the Graph-NLPCA, 96078.

Table 1. Root-mean-square error (RMSE) on the estimates for different proportions of missing data.

Missing data ratio	0.1	0.2	0.3	0.4	0.5
Graph-NLPCA	0.0026	0.0040	0.0051	0.0061	0.0077
NLPCA	0.0043	0.0043	0.0045	0.0048	0.0071

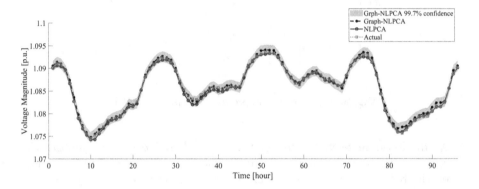

Fig. 3. Example of voltage estimation given only data of power and renewable generation.

Table 1 summarised the root-mean-square error (RMSE) obtained for a proportion of 10% to 50% of missing data. Based on the results, it can be concluded that the factorization enforced by the proposed Graph-NLPCA model does not affect negatively the modelling accuracy with respect to the case where no assumptions about variable dependency is made.

Figure 3 shows example results where all the node voltages are estimated given only the power injections, which is a typical use-case for power-flow simulation models. Again, no noticeable difference between the Graph-NLPCA and NLPCA model is observed.

3.3   Detecting Erroneous Data

As outlined in Sect. 1, a desired feature of a data-driven power system model is the ability to perform anomaly detection, also referred to as bad data analysis [3]. A common example is the installation of new renewable generation plant, which may come on-line well before any metering data are made available.

An experiment was designed by modifying the validation data based on a power-flow simulation after the solar generation data for one of the points was doubled. The measurements for the particular solar generation point, however, was kept at the same level as in the training data. Figure 4 shows how the resulting estimate is consistently higher than the measurement and a statistical test can be designed to detect the anomaly (a Z-test of the residuals, with respect to the standard deviation produced by the model, gives a probability of about 0.9999 that the estimate is larger than the data).

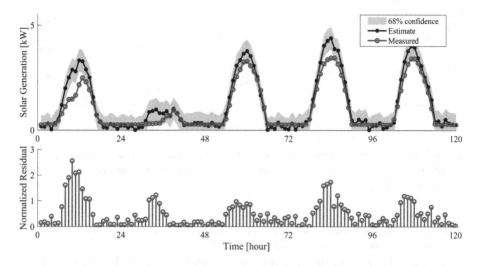

Fig. 4. Example of detection of unaccounted solar generation.

3.4 Scalability

In order to demonstrate the excellent scalability properties of the proposed graphical model, a series of experiments was run where the dimensionality of the problem was increased from the 10 graph partitions (375 variables) used for validation in Sect. 3.2, to 50 (1416), 100 (2576), 150 (3719) and 200 (4643).

Figure 5 compares the RMSE estimation accuracy obtained with Graph-NLPCA and NLPCA when 10% of the data was missing on the validation set. While the two models behave similarly up to around 2000 variables, the performance of the centralised approach based on NLPCA starts deteriorating thereafter as the dimensionality of the problem increases. This is to be expected as the number of parameters, also shown in Fig. 5, grows exponentially thus negatively affecting the training phase where only limited data are available. The proposed Graph-NLPCA, on the other hand, maintains a consistent level of error accuracy throughout.

4 Conclusion

An innovative approach to data-driven power system modelling based on probabilistic graphs was introduced. By providing a flexible computational framework where custom non-linear, localised models of the joint density can be combined

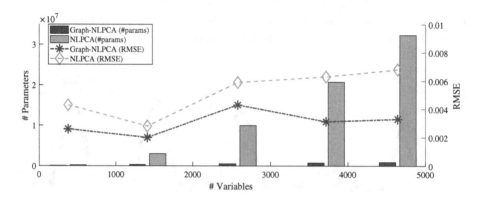

Fig. 5. Scalability of number of parameters and modelling accuracy with respect to variable dimensionality. Accuracy is based on the estimation error when 10% of the data are missing.

to model arbitrarily large and complex systems, the method is well suited to handle the complexity and scale of the electrical grid.

A specific instance of the graphical model, where the joint densities are modelled with a neural-network approach to NLPCA, was evaluated on data imputation and anomaly detection applied to a publicly available data set representative of the European transmission grid. The graph structure is based on a simple heuristics applied to the grid topology and allows for excellent scalability properties, where the model accuracy is not negatively affected by growing dimensionality when only limited training data are available, as opposed to a centralised model.

Many other possibilities exist to model the individual factor nodes and different models could even be combined in the same graph. Identifying the most suitable model (or set of models) in the context of power systems will be an important topic for future investigation. A naive approach was considered to initialize the structure of the graphical model and applying more formal data-driven techniques will also be scope for further studies. Finally, expanding the model to include time dependencies, such to cover a wider range of prediction tasks and to increase robustness to missing data, will also be an interesting research direction.

Acknowledgements. This research has received funding from the European Research Council under the European Union's Horizon 2020 research and innovation programme (grant agreement no. 731232). The author would like to thank Sean McKenna and Bradley Eck, from IBM Research Ireland, for pointing to the data set used in this research, and Michele Berlingherio, from IBM Research Ireland, for providing excellent feedback on the manuscript.

References

1. Abadi, M., et al.: Tensorflow: a system for large-scale machine learning. In: Proceedings of the 12th USENIX Symposium on Operating Systems Design and Implementation (OSDI) (2016)
2. Abur, A., Exposito, A.G.: Detecting multiple solutions in state estimation in the presence of current magnitude measurements. IEEE Trans. Power Syst. 12(1), 370–375 (1997). https://doi.org/10.1109/59.575721
3. Abur, A., Exposito, G.A.: Power System State Estimation: Theory and Implementation. CRC Press, Boca Raton (2004)
4. Barber, D.: Bayesian Reasoning and Machine Learning. Cambridge University Press, Cambridge (2012)
5. Collier, S.E.: The emerging enernet: convergence of the smart grid with the Internet of Things. IEEE Ind. Appl. Mag. 23(2) (2017)
6. Cosovic, M., Vukobratovic, D.: Distributed Gauss-Newton method for AC state estimation: a belief propagation approach. In: IEEE International Conference on Smart Grid Communications (SmartGridComm) (2016)
7. Cosovic, M., Vukobratovic, D.: State estimation in electric power systems using belief propagation: an extended DC model. In: IEEE International Workshop on Signal Processing Advances in Wireless Communications (SPAWC) (2016)
8. Fiedler, M.: Laplacian of graphs and algebraic connectivity" and combinatorics and graph theory. 25(1), 57–70. Banach Center Publications (1989)
9. Frey, B.J., Jojic, N.: A comparison of algorithms for inference and learning in probabilistic graphical models. IEEE Trans. Pattern Anal. Mach. Intell. 27(9), 1392–1416 (2005)
10. Fusco, F., Thirupathi, S., Gormally, R.: Power systems data fusion based on belief propagation. In: Proceedings of the IEEE PES Innovative Smart Grid Technologies (ISGT) Conference Europe (2017)
11. Honkela, A., Valpola, H.: Unsupervised variational Bayesian learning of nonlinear models. In: Proceedings of the Advances in Neural Information Processing Systems (NIPS) 17 (2004)
12. Hu, Y., Kuh, A., Kavcic, A., Yang, T.: A belief propagation based power distribution system state estimator. IEEE Comput. Intell. Mag. 6, 36–46 (2011)
13. Jensen, T.V., Pinson, P.: Re-Europe and a large-scale dataset for modeling a highly renewable european electricity system. Sci. Data 4, 170–175 (2017)
14. Josz, C., Fliscounakis, S., Maeght, J., Panciatici, P.: AC power flow data in MATPOWER and QCQP format: iTesla and RTE snapshots and and PEGASE. arXiv:1603.01533 (2016)
15. Kingma, D.P., Welling, M.: Auto-encoding variational Bayes. In: Proceedings of the Sixth International Conference on Learning Representations (ICLR) (2014)
16. Koller, D., Friedman, N.: Probabilistic Graphical Models. MIT Press, Cambridge (2009)
17. Luttinen, J., Llin, A.: Variational Gaussian-process factor analysis for modeling spatio-temporal data. In: Advances in Neural Information Processing Systems (NIPS) 22 (2009)
18. Nguyen, M.H., De la Torre, F.: Robust kernel principal component analysis. In: Advances in Neural Information Processing Systems (NIPS) 22 (2009)
19. Sanguinetti, G., Lawrence, N.D.: Missing data in Kernel PCA. In: Fürnkranz, J., Scheffer, T., Spiliopoulou, M. (eds.) ECML 2006. LNCS (LNAI), vol. 4212, pp. 751–758. Springer, Heidelberg (2006). https://doi.org/10.1007/11871842_76

20. Scholz, M., Kaplan, F., Guy, C.L., Kopka, J., Selbig, J.: Gene expression non-linear PCA : a missing data approach. Bioinformatics **21**(20), 3887–3895 (2005). https://doi.org/10.1093/bioinformatics/bti634
21. Zimmerman, R.D., Murillo-Sánchez, C.E., Thomas, R.J.: Matpower: steady-state operations and planning and analysis tools for power systems research and education. IEEE Trans. Power Syst. **26**(1), 12–19 (2011)

Renewable Energy Integration: Bayesian Networks for Probabilistic State Estimation

Ole J. Mengshoel$^{(\boxtimes)}$, Priya K. Sundararajan, Erik Reed, Dongzhen Piao, and Briana Johnson

Carnegie Mellon University, Pittsburgh, USA
ole.mengshoel@sv.cmu.edu

Abstract. Increased availability of renewable energy sources, along with novel techniques for power flow control, open up a broad range of interesting challenges and opportunities in power flow optimization. This promises reduced power generation costs through better integration of renewable energy generators into the Smart Grid. Unfortunately, renewable generators are fundamentally variable and uncertain. This uncertainty motivates our study of probabilistic state estimation techniques in this paper. Specifically, we use probabilistic graphical models in the form of Bayesian networks to compute probabilities of power system states, thus enabling improved power flow control. Key differences between our probabilistic state estimation results as reported in this paper and similar previous efforts include: our use of Bayesian probabilistic but exact (rather than Monte Carlo) state estimation techniques; auto-generation of Bayesian networks for probabilistic state estimation; integration with corrective Security-Constrained Optimal Power Flow; and application to Distributed Flexible AC Transmission Systems. We present novel models and algorithms for probabilistic state estimation using auto-generated Bayesian networks compiled to junction trees, and report on experimental results that illustrate the scalability of our methods.

Keywords: Renewable energy · Optimal power flow
Power flow control · Smart wires · Transmission grid
Probabilistic state estimation · Bayesian networks · Junction trees
Exact inference

1 Introduction

As part of the emerging Smart Grid, many improvements to hardware and software for power transmission, power distribution, energy management, sensors, communication, and computation are taking place. These improvements include novel techniques for power flow control by means of Distributed Flexible AC Transmission Systems (D-FACTS) [15]. Such better techniques for power flow control, along with increases in renewable energy generation, create many opportunities and challenges in power flow optimization [20,33]. The development

© Springer Nature Switzerland AG 2018
W. L. Woon et al. (Eds.): DARE 2018, LNAI 11325, pp. 63–82, 2018.
https://doi.org/10.1007/978-3-030-04303-2_5

of D-FACTS and supporting technologies promises increased utilization of the transmission grid as well as reduced generation costs through better integration of renewable energy generators into the grid.

In support of the integration of D-FACTS devices into the emerging Smart Grid, this paper presents novel results on models and algorithms for probabilistic state estimation using Bayesian networks (BNs) [37]. BNs, which model multivariate probability distributions by means of directed acyclic graphs, are a key artificial intelligence technique. BNs have several benefits—they enable mathematical modelling, visualization, efficient state estimation inference algorithms, and machine learning—and have proven themselves in a broad range of power system and other applications [8,12,18,19,25,29,39,43,44,50].

Using BNs, we construct mathematical models of power systems including generation with renewable energy sources. These BN models are then used to efficiently compute posterior marginal distributions that assist in evaluating the impact of D-FACTS on power flow in the grid. Key characteristics of our probabilistic state estimation (PSE) approach include: (i) auto-generation of BNs used for PSE; (ii) exact computation of posterior marginals (not Monte Carlo simulation) via compilation of a BN to a secondary data structure, a junction tree; and (iii) integration with Security-Constrained Optimal Power Flow (corrective SCOPF) computation as part of a Smart Grid architecture with D-FACTS. We illustrate the benefit and scalability of our data-driven approach in simulations with up to 118-bus systems.

The rest of this paper is structured as follows. In Sect. 2, we provide background on BNs and some Smart Grid techniques. Section 3 discusses the transmission and distribution system context of our techniques, including integration with corrective SCOPF and D-FACTS. In Sect. 4, we present our probabilistic state estimation techniques, including how BNs are constructed. Section 5 discusses experimental results with IEEE test systems containing up to 118 buses.

2   Background

While power flow analysis has mostly been performed in a deterministic fashion [1], probabilistic load flow analysis was discussed already in the 1970s [6]. This section discusses related work and background for our probabilistic state estimation approach. We briefly review probabilistic and Bayesian methods, introduce a few fundamental Smart Grid concepts, and discuss prior power systems applications of probabilistic modeling and computation. Finally, we highlight similarities and differences between our effort and previous work.

2.1   Probabilistic Graphical Models and Bayesian Networks

Principles. Bayesian networks (BNs) represent multivariate probability distributions and are used for reasoning and learning under uncertainty [37]. Similar to other probabilistic graphical models (PGMs), probability theory and graph theory form the foundation of BNs. Random variables are represented as nodes in a directed acyclic graph (DAG), while conditional dependencies and independencies are reflected in the DAG's edges. A BN is an especially useful model if a sparse graph model of the underlying phenomena can be developed. Fortunately, graph sparsity is natural in models of electrical power systems, especially transmission systems, due to the high cost of their build-out and operation.

Notation, Terminology, and Definitions. The following notation will be used throughout this paper: Let X be the random variables in a BN. A BN factorizes the joint distribution $\Pr(X)$ via a DAG. The random variables are represented as DAG nodes, with W being the DAG edges. Finally, let P be conditional probability distributions (CPDs), one per node $X \in X$. CPDs are often conditional probability tables (CPTs). Now, a BN β is defined by the tuple (X, W, P) [37].

A BN $\beta = (X, W, P)$ enables different probabilistic queries to be formulated and then answered by efficient inference algorithms. We identify $E \subset X$ as evidence or input nodes. The notation e is used to denote the evidence or input, which assigns to each node $E \in E$ a value. An inference algorithms assumes that the nodes in E are "clamped" to values e, and computes or approximates the posterior $\Pr(Y \mid e)$, where $X = Y \cup E$ and $Y \cap E = \emptyset$. The nodes Y are denoted the non-evidence or output nodes. An explanation y is an instantiation of all nodes in Y. Computation of a most probable explanation (MPE) amounts to finding y^*, an explanation y among all possible explanations, namely one that maximizes $\Pr(y \mid e)$: MPE(e). Computation of marginals (or beliefs) amounts to inferring the posterior probabilities over one or more query nodes $Q \subseteq Y$, specifically BEL(Q, e) where $Q \in Q$. Marginals can be used to compute most likely values by picking, in BEL(Q, e), a most likely state.

Computation Generally. Many BN inference problems are computationally hard (NP-hard or worse) in the general case [9,40,46]. However, algorithms that are efficient in practice have been developed, both exact algorithms [7,11,14,26, 45] and inexact algorithms [21,24,32].

Exact algorithms include junction tree propagation [3,23,26,45], conditioning [10,37], variable elimination [14,51], and arithmetic circuit evaluation [7,11]. Inexact (or approximate) algorithms include stochastic simulation, Markov-chain Monte Carlo (MCMC), variational inference [22], and stochastic local search (SLS) algorithms. SLS algorithms have been used to compute MPEs [21,24,32] as well as MAPs [36] in Bayesian networks.

Computation with Junction Trees. To address real-time reasoning challenges often associated with power systems, we can compile BNs into junction trees (JTs) or arithmetic circuits (ACs) [29,39]. Both these data structures, even when implemented in software, support real-time settings including Smart Grids by being predictable, fast, and exact. A well-established and popular method for BN inference, useful for both MPE and marginal computation over $\Pr(Y \mid e)$, is to use JTs compiled from BNs [3,23,26,45]. In brief, the JT method works as follows. The DAG structure of the BN is compiled, in an off-line step, into a tree structure. The tree, known as a junction tree, consists of cliques and separators. Each clique and each separator is a supernode in the sense that it consists of multiple nodes from the original BN. The tree is structured in such a way that when evidence e is propagated, on-line, exact marginal or MPE computation is achieved. (The details for marginal versus MPE computation are slightly different.) In this work, we are studying the use of JTs in the Smart Grid setting, including their Achilles heel of substantial growth in JT size.

2.2   Power Flow, Optimization, Security, and Smart Wires

The goal of power flow control is to steer how electrical power flows in transmission lines. Specifically, power flow control amounts to steering power from generators, both traditional and renewable, via a desired path to the loads. Optimizing power flow, or the problem of optimal power flow (OPF), plays a key role in power systems. In essence, the OPF problem is to minimize generation cost to meet a certain demand, subject to various constraints [2,20,33]. OPF is a central issue in both traditional and renewable power generation.

Security, in the context of power flow optimization, means that no constraints are violated in normal operation or under certain failure scenarios. For example, $N-1$ security means that the transmission grid is operational under normal operational conditions and even under conditions where there is a failure of one element in the transmission system. Security levels of $N-1$ (one failure) or $N-2$ (two failures) are common in the power industry.

A mathematical optimization problem that minimizes generation cost can be formulated to formalize secure operation, defining the concepts of Security-Constrained Optimal Power Flow (SCOPF). There is a distinction between preventive SCOPF and corrective SCOPF [20,33]. Corrective SCOPF, which is our main focus in this paper, amounts to taking actions to remove constraint violations to ensure security [33].

Small changes in impedance values can have significant impact on the power flow in transmission lines. Interestingly, such impedance changes can be achieved by small modules that can easily be attached to existing power lines. In other words, these modules augment rather than replace the current power transmission infrastructure. These modules can easily be added in a distributed fashion across a transmission grid and form one of the main concepts of distributed FACTS (D-FACTS).

In essence, D-FACTS and smart wires make up a solution to transmission congestion and reliability issues [15]. Alternative solutions, such as deploying new power lines or Flexible AC Transmission Systems (FACTS), have turned out to be not so reliable and less cost-efficient.

2.3   Probabilistic Graphical Models in Power Systems

PGMs and BNs have been successful in a wide range of applications. Most relevant to this work are power systems applications in state estimation [18,19] and model-based fault diagnosis [8,27,29,50]. One example power system application is a micro-grid known as the Advanced Diagnostic and Prognostic Testbed (ADAPT) [38], located at the NASA Ames Research Center. For the ADAPT micro-grid, BNs have been used to jointly perform state estimation and diagnosis of a broad range of power system faults, including offset faults, drift faults, and intermittent faults [29,39]. The diagnosis approach is based on explicitly representing the "health" of individual power system components and sensors including batteries, relays, voltage sensors, and loads in nodes of the BNs, and then use BN inference to diagnose which components are "healthy" or "unhealthy" [29,30,39].

BNs have also been used to diagnose faults in large power generating stations [47] and other aspects of power generation and distribution [8,50]. Other PGMs, in particular factor graphs, have been used to perform real-time state estimation in micro-grids [18,19].

2.4   Similarities and Differences

The concept of probabilistic state estimation (PSE) is well-established [2,6,42, 49], and serves as the foundation for this work. Within the area of PSE, how is the work reported in this paper different from previous work (that we know)?

In brief, there is a difference between related previous work and this work in our focus on (i) exact Bayesian inference methods with BNs and JTs for (ii) power flow control for transmission systems in the context of (iii) security-constrained optimal power flow. Let us now discuss these differences in some more detail.

Regarding (i): There is previous research on the application of Bayesian methods including PGMs to power systems, but in many cases these efforts resort to approximate inference rather than exact inference when computing $\Pr(Y \mid e)$ [42,49]. While approximate methods may be required in certain applications, such methods introduce the question of the accuracy of the approximation $\widehat{\Pr}(Y \mid e)$. This question is, by definition, not raised for exact methods. On the other hand, the complexity and scalability of exact JT inference can be a concern. Intuitively, as the connectivity of the BN increases, the number of nodes in one or more JT cliques increases dramatically. Thus, since clique size is exponential in the number of nodes in the clique, the impact on memory consumption and inference time can be severe [28,31].

Regarding (ii) and (iii): Compared to previous research that uses exact inference in PGMs for power system analysis [29,39], our focus on power flow in transmission systems, rather than fault detection and diagnosis, makes this work different. For example, previous work studied the ADAPT micro-grid. ADAPT was modeled using BNs, which were compiled to arithmetic circuits (ACs) and junction trees (JTs), which were then used for exact inference [29,39].[1] However,

---

[1] In experiments with different ADAPT configurations and BNs, strong detection and diagnostic performance was achieved. In international diagnostics competitions arranged in 2009 and 2010, this approach had the best results in three out of four competitions in the track that used ADAPT. Please see https://c3.nasa.gov/dashlink/projects/36/ and https://c3.nasa.gov/dashlink/projects/36/ for further details.

such previous PGM research on fault detection and diagnosis has not considered power flow optimization, D-FACTS, or corrective SCOPF.

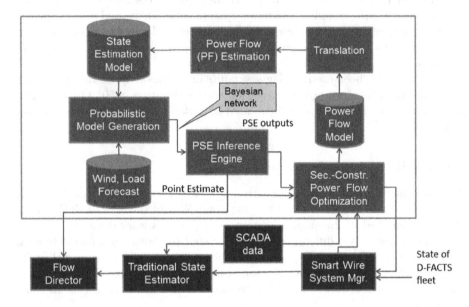

Fig. 1. Smart Grid system architecture, including the components Probabilistic Model Generation, Bayesian network, Probabilistic State Estimation (PSE) Inference Engine (using BNT), Security-Constrained Power Flow Optimization (SCOPF using Tomlab), Power Flow (PF) Estimation (using MatPower), Flow Director, and Smart Wire system Manager (using D-FACTS).

3   Smart Grid System Architecture

This section discusses probabilistic state estimation (PSE) techniques in a Smart Grid context, see Fig. 1. We are especially focused on the interaction between PSE, D-FACTS, and OPF, including security-constrained optimal power flow (corrective SCOPF).

While some information flows are left out in Fig. 1, to improve readability, note that both corrective SCOPF and PSE use many electrical power system parameters. These parameters include generator parameters (generator types, costs, and production limits) and power line parameters (line impedances and limits). In addition, there are system loads as well as outputs of generators that rely on renewable energy sources, for example outputs of wind power plants.

3.1   Role of Smart Wires

D-FACTS [15] is used to control the impedance (mainly reactance) of a power line, which in turn can increase the loading margin of the transmission grid by pushing power flow to lines with spare capacity. In Fig. 1 a centralized Smart Wire System Manager is used to access and control the D-FACTS devices.

The corrective SCOPF component optimizes the parameters used by D-FACTS power flow control, as well as the parameters for power generation by the non-renewable energy sources [20,33].[2] The objective of the corrective SCOPF model is to obtain minimum generation cost, taking advantage of the integration of D-FACTS into the power transmission grid, while maintaining system security.

3.2 Role of Security-Constrained Optimal Power Flow (SCOPF)

The input parameters of corrective SCOPF consist of the network structure, the parameters for the line limits, types of generators, cost function parameters associated with the generators, etc. The output parameters of corrective SCOPF consist of the optimized generator power dispatches and the D-FACTS line reactances. SCOPF computes an optimal power flow model that minimizes generation cost, while at the same time maintaining system security [33].

Power Flow (PF) estimation is performed using these parameters, resulting in the State Estimation Model in Fig. 1.[3] In fact, we convert both the input and output parameters of corrective SCOPF (which are stored in the Power Flow Model as shown in Fig. 1) into PF estimation format.

3.3 Role of Probabilistic State Estimation (PSE)

A Smart Grid BN model $\beta = (X, W, P)$ is used to model the power system for the purpose of state estimation. Specifically, we form a joint distribution $\Pr(X)$, from which many questions of interest can be answered. We are interested in the posterior $\Pr(Y \mid e)$, where e is evidence or input such as commands and sensor readings. This BN modeling is further discussed in Sect. 4.1.

The Probabilistic Model Generation process in Fig. 1 takes the PF output, stored in the State Estimation Model in Fig. 1, as well as forecast data from external sources (e.g., for wind, solar) to generate a BN model $\beta = (X, W, P)$. To generate β we use outputs of corrective SCOPF, namely the optimized production limits for generators and transmission line reactances, as input (via the Power Flow Model in Fig. 1). This model generation process is further discussed in Sect. 4.2.

The PSE Inference Engine in Fig. 1 uses the BN model β, in the form of a JT compiled from the BN, to compute a posterior estimate $\Pr(Y \mid e)$ of the power system's state. This estimate $\Pr(Y \mid e)$ can include non-traditional state variables among Y, such as the health of components and sensors [29,30,39] in the transmission grid. This inference process is further discussed in Sect. 4.3.

---

[2] We assume power flow control devices such as smart wires; please see http://www. smartwiregrid.com/ for details.

[3] Details on distributed control techniques for solving DC OPF problems when transmission lines are instrumented with D-FACTS have been presented by Mohammadi, Hug, and Kar [34].

3.4 Discussion

Key distinguishing factors of our Smart Grid approach, as discussed here in Sect. 3 and summarized in Fig. 1, include the following. Traditional energy management systems (EMSs) do not compute probabilities, while our PSE Inference Engine does. Specifically, the SCOPF algorithm provides inputs that PSE uses to form a Smart Grid BN $\beta = (X, W, P)$ and then using β compute the posterior estimate $\Pr(Y \mid e)$, where $Y \subseteq X$, of the transmission system's state. While the details of corrective SCOPF are beyond the scope of this paper, we also note this: Unlike our corrective SCOPF algorithm, traditional corrective SCOPF algorithms do not take into account the operation and control of D-FACTS devices by means of $\Pr(Y \mid e)$ as computed by the PSE Inference Engine.

4 Probabilistic State Estimation Techniques

We now present our probabilistic state estimation models, in the form of BNs. We also discuss how they are constructed and then used for probabilistic state estimation.

4.1 Probabilistic State Estimation Model

Our approach takes as a starting point bus and line data in the form of a graph $G = (U, B)$, where U are nodes (representing buses) and B are edges (representing branches).The result is a BN $\beta = (X, W, P)$, where X are nodes, W are edges, and P are CPDs. A joint distribution $\Pr(X)$ is induced by β according to the DAG's structure:

$$\Pr(X) = \prod_{i=1}^{n} \Pr(X_i \mid \mathrm{pa}(X_i)),$$

where $\mathrm{pa}(X_i)$ denotes the parents of X_i in the DAG and $\Pr(X_i \mid \mathrm{pa}(X_i))$ denotes node X_i's CPD.

We create a BN in which power system buses U become BN leaf nodes in X and branches (or lines) B become BN root nodes in X. Associated with a bus, there can also be BN nodes for a load and a generator (none, one, or both of which may exist for a given bus). Both the buses and the branches are "first-class citizens"and are represented as random variables in X. Figure 2 shows an example BN for a small 4-bus power system.

We have explored different BN approaches to represent the impact of D-FACTS. Currently, the impact of D-FACTS is reflected in the parameters P of BN nodes. In other words, BN node parameters vary, depending on the D-FACTS settings as determined by corrective SCOPF. This is sufficient as long as we do not need to reason about the state of the D-FACTS devices with the BN itself.

While corrective SCOPF and MatPower are based on deterministic models, PSE uses probabilistic models with probability parameters. Whether D-FACTS

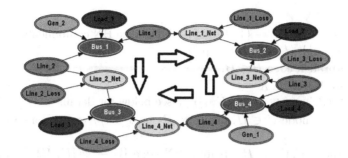

Fig. 2. The DAG part of a 4-bus BN model $\beta = (\boldsymbol{X}, \boldsymbol{W}, \boldsymbol{P})$ is shown, with nodes in $\boldsymbol{X}$ color-coded by type. Generator nodes are grey, load nodes are red, bus nodes are blue, line nodes are lime green, line loss nodes are orange, and line net nodes are yellow. The small black edges indicate conditional distributions associated with the BN nodes, while the four large edges in the center indicate power flow, away from the two generators, Gen_1 and Gen_2. (Color figure online)

is on or off impacts the parameters of the BNs but not their structure. See Fig. 2 for an example BN $\beta = (\boldsymbol{X}, \boldsymbol{W}, \boldsymbol{P})$; probabilities $\boldsymbol{P}$ are for simplicity suppressed. In this figure, we first note that power flow is according to the four large arrows in the center, not the smaller black arrows representing BN edges $\boldsymbol{W}$. Second, we note that the BN enables important types of uncertainties in power systems to be represented (*e.g.*, uncertainty of renewable energy generation, line loss uncertainty, and load uncertainty).

Necessary parameters to generate the BN in Fig. 2 include Gaussian means and variances of the buses, loads, lines, generators, and line losses. The mean values associated with the Gaussian random variables were taken from the output of corrective SCOPF and MatPower (see Sects. 3 and 5). The standard deviations associated with the Gaussian random variables were determined based on historical power systems data (see Sect. 5). While the BN in Fig. 2 contains Gaussian nodes only, the algorithm discussed in Sect. 4.2 can also incorporate multinomial (discrete) nodes.

4.2   Probabilistic State Estimation Model Construction

The CREATESTATEESTIMATOR($\boldsymbol{G}$, λ, γ) algorithm below takes as input a graph $\boldsymbol{G} = (\boldsymbol{U}, \boldsymbol{B})$ and outputs a probabilistic state estimator, a BN β.[4] The algorithm also takes as input a load map λ and generator map γ. This algorithm formalizes the Probabilistic Model Generation process in Fig. 1, and works as follows:

---

[4] The graph definition $\boldsymbol{G} = (\boldsymbol{U}, \boldsymbol{B})$ is found, for example, in an output file produced by the MatPower software, which is used in our experiments in Sect. 5. The MatPower file contains two tables called Bus Data and Branch Data. These tables contain the graph $\boldsymbol{G} = (\boldsymbol{U}, \boldsymbol{B})$ discussed here along with other data utilized by CREATESTATEESTIMATOR.

1. Create $\beta = (\boldsymbol{X}, \boldsymbol{W}, \boldsymbol{P})$
   (a) Let $\boldsymbol{X} \leftarrow \{\}$, $\boldsymbol{W} \leftarrow \{\}$, $\boldsymbol{P} \leftarrow \{\}$ // initialize BN components
   (b) Let $\Phi \leftarrow \{\}$ // initialize map
2. For each branch edge $B \in \boldsymbol{B}$:
   (a) Create BN node K
   (b) CREATEFRAGMENT(B, K, β) // see below for details
3. For each bus node $U \in \boldsymbol{U}$, create a BN node Y // create constraint (leaf) node
   (a) For each $\{U, V\} \in \boldsymbol{B}$, create a BN edge $W \leftarrow (\Phi(\{U, V\}), Y)$ and put $\boldsymbol{W} \leftarrow \boldsymbol{W} \cup \{W\}$
   (b) If U has a load $\lambda(U) \neq$ NIL,
      i. Create BN node K // power state node
      ii. CREATEFRAGMENT(U, K, β) // see below for details
      iii. Create a BN edge $W \leftarrow (X, Y)$
      iv. Let $\boldsymbol{W} \leftarrow \boldsymbol{W} \cup \{W\}$
   (c) If U has a generator $\gamma(U) \neq$ NIL
      i. Create BN node K // power state node
      ii. CREATEFRAGMENT(U, X, H, S, β) // see below for details
      iii. Create a BN edge $W \leftarrow (X, Y)$
      iv. Let $\boldsymbol{W} \leftarrow \boldsymbol{W} \cup \{W\}$
4. For each node $X \in \boldsymbol{X}$:
   (a) If X is a bus node, create bus node CPD P
   (b) If X is a branch node, create line node CPD P
   (c) If X is a branch node, create load node CPD P
   (d) Let $\boldsymbol{P} \leftarrow \boldsymbol{P} \cup \{P\}$
5. Return $\beta = (\boldsymbol{X}, \boldsymbol{W}, \boldsymbol{P})$

The above CREATESTATEESTIMATOR algorithm works by creating a bipartite-like BN (see Fig. 2). In such a BN, bus BN nodes (colored blue in Fig. 2) are leaf nodes and line BN nodes (colored lime green in Fig. 2), along with BN nodes for a load and a generator (none, one, or both of which may exist for a given bus), are root nodes.

We now consider the node CPDs $\boldsymbol{P}$. The case of discrete multinomial CPDs, for health nodes, is straightforward. The more interesting case is continuous Gaussian CPDs, which can be found in hybrid BNs, where nodes can be either discrete or continuous.[5] Consider a continuous node Z. Its continuous parents (if any) are called $\boldsymbol{C}$ and its discrete parents (if any) are called $\boldsymbol{D}$. The distribution on Z is defined as follows:

- No parents: $Z \sim N(\mu, \boldsymbol{\Sigma})$.
- Continuous parents: $Z \mid \boldsymbol{C} = \boldsymbol{c} \sim N(\mu + \boldsymbol{W}\boldsymbol{c}, \boldsymbol{\Sigma})$.
- Discrete parents: $Z \mid \boldsymbol{D} = \boldsymbol{d} \sim N(\mu(:, \boldsymbol{d}), \boldsymbol{\Sigma}(:, :, \boldsymbol{d}))$.

---

[5] The junction tree algorithm works for certain hybrid BNs, namely those in which continuous (Gaussian) nodes can have both continuous and discrete parents, but discrete (multinomial) nodes can only have discrete parents.

– Continuous and discrete parents: $Z \mid C = c, D = d \sim N(\mu(:, d) + W(:, :, d) \cdot c, \Sigma(:, :, d))$.

Here, $N(\mu, \Sigma)$ denotes a Gaussian distribution with mean μ and covariance Σ. Let $|Z|$, $|C|$, and $|D|$ denote the sizes of Z, C, and D respectively. If there are no continuous parents, $|C| = 0$; if there is more than one, then $|C| =$ the sum of their sizes. Then μ is a $|Z| * |D|$ vector, Σ is a $|Z| * |Z| * |D|$ positive semi-definite matrix, and W is a $|Z| * |C| * |D|$ regression (weight) matrix. The situation is similar for the case of discrete parents.

The CREATEFRAGMENT(B, K, (X, W, P)) algorithm is called by CREATESTATEESTIMATOR. CREATEFRAGMENT takes as input an edge B, a power state BN node K, and a BN $\beta = (X, W, P)$, and works as follows:

1. Let $X \leftarrow X \cup \{K\}$ // add power state node as root
2. Let $\Phi \leftarrow \Phi \cup (B, X)$ // map[6] from G to β
3. Create a BN edge $W \leftarrow (X, S)$ and put $W \leftarrow W \cup \{W\}$
4. Create a BN edge $W \leftarrow (H, S)$ and put $W \leftarrow W \cup \{W\}$

A key difference between the power network G and the BN β is that the former has transmission lines (or branches) represented as edges, while in the BN the transmission lines are "first-class citizens" and are represented as random variables. This allows us to compute posteriors that reflect the state of transmission lines. Among BNs, we distinguish between continuous and hybrid (continuous plus discrete) BN models. We are in general interested in hybrid BN models of Smart Grids, as discussed above. However, our main focus is on continuous BNs (see Fig. 2) in the experiments in Sect. 5. This is due to a lack of data for discrete nodes such as health, command, and sensor nodes.

4.3 Probabilistic State Estimation Process

The probabilistic state estimation algorithm STATEESTIMATION takes as input a constructed BN $\beta = (X, W, P)$ along with other inputs (see Sect. 4.2). This algorithm corresponds to the PSE Inference Engine process in Fig. 1. We now briefly describe the on-line STATEESTIMATION process along with its inputs and outputs.

Along with β, PSE input includes evidence $e = e_F \cup e_V$, where e_F is the fixed evidence, e_V is the variable evidence, and $e_F \cap e_V = \emptyset$. Sensor readings and commands are part of e_V while physical constraints and laws, such as power conservation, are part of e_F. A bus node X is used as a fixed evidence (or constraint) node, where $X = 0$ is part of e_F and used to represent a power conservation law. Since it is part of the evidence and is a leaf node, X induces in $\Pr(Y \mid e)$ a conditional dependency between all of its parent nodes which in physical terms "correlates" the power flow on the branches connected to the bus.

---

[6] Both relational and functional notation is used for Φ. In other words, we say (relationally) $(B, X) \in \Phi$ when Φ maps (functionally) from B to X.

The outputs from probabilistic state estimation are based on the posterior distribution $\Pr(Y \mid e)$. From the posterior distribution $\Pr(Y \mid e)$, after compiling the BN to a JT, we obtain marginal power state estimates for the loads, generators, and line nodes. Each of these random variables have power limits associated with them. The state estimates for the loads and the renewable generators are used as input to corrective SCOPF, which outputs the optimal values for the conventional generators.[7] The state estimates of the line nodes are used to find the impact of D-FACTS devices on the Smart Grid. The posterior distribution $\Pr(Y \mid e)$ can also used to find out which health nodes, if any, are in non-healthy states, thus suggesting that one or more failures exist in the transmission grid or in the D-FACTS devices.

5 Experimental Results

In this experimental section, we first describe the methods and data. Next, we focus on experimental results with the PSE methods. Simulation results to demonstrate the potential benefit and scalability results to study the suitability of exact BN inference by means of JTs are both discussed.

5.1 Methods and Data

Methods. MatPower was used in experiments to compute power flow data [55]. MatPower is a MATLAB package for simulating and solving power flow and optimal power flow problems. BNT was also employed. BNT [35] is an open source MATLAB package that can be used for experimenting with hybrid BNs. In our case, a BN is generated from power flow models output by MatPower. Our PSE methods use BNT's JT method to do exact inference.

Tomlab [17], an optimization library for MATLAB, is used to run corrective SCOPF. Simulated D-FACTS power flow control devices are attached to the transmission lines. These devices are used to control the reactances of the power lines, resulting in varying power flows, reflecting the outputs of corrective SCOPF (see Sect. 3).

Data. We study transmission systems with 4 to 118 buses; see Table 1. For example, the IEEE 39-bus system used in experiments is well-known as the 10-machine New-England Power System. In the experiment, the transmission systems data come from MatPower, which takes its data from the literature [5].[8] After running the optimal power flow algorithm in MatPower, the loads of buses and generators are extracted for state estimation purposes.

In our renewable variant of the IEEE 39-bus test system, there is power generated by renewable energy sources, specifically wind power plants. This variant of the IEEE 39-bus test system, used in some of the experiments, continues to

---

[7] For example, the IEEE 39-bus as discussed elsewhere in this paper has a total of 10 generators (8 conventional generators and 2 wind power generators) and 19 loads. In this case, there are 21 inputs to corrective SCOPF from PSE.

[8] The data can be found in the file `case39.m` in the MatPower package.

Table 1. Evaluation of test systems, with different number of buses, when converted to Smart Grid BNs. All computation times (convert, compile, and inference) refer to the use of junction trees (JTs) compiled from Bayesian networks (BNs) and are in seconds.

System name	Buses $|U|$	Lines $|B|$	Loads	Generators	Nodes $|X|$	Convert Time (s)	Compile Time (s)	Inference Time (s)
case4gs	4	4	4	2	22	0.02	0.03	0.03
case14	14	20	11	5	90	0.10	0.28	0.11
case39	39	46	21	10	208	0.22	1.31	0.25
case118	118	186	99	54	829	1.23	60.97	1.16

have 10 generators. However, two of the generators (Generator 3 and Generator 8) are considered wind power plants with uncertain outputs. In the experiments, historical data were used as a basis for modeling this variability. In these experiments, the corrective SCOPF component [33] optimizes the parameters used by D-FACTS controllers.

We also use a Bonneville Power Authority (BPA) SCADA system dataset in our experiments.[9] This time series dataset is based on samples collected every 2 s. Using this dataset, we calculated the variances needed to set BN parameters $\Pr(X \mid \mathrm{pa}(X))$, for the relevant $X \in \boldsymbol{X}$. These variances complement the means coming from corrective SCOPF or MatPower output; together they were used as Gaussian parametric input to a BN $\boldsymbol{\beta} = (\boldsymbol{X}, \boldsymbol{W}, \boldsymbol{P})$.

5.2 Simulation Results

The usage of D-FACTS helps to control the impedance in the transmission line and can enable higher reliability and utilization of power systems. Varying D-FACTS settings may have positive or negative impacts on a power transmission system when viewed from a probabilistic perspective. To better understand such impacts, we integrated the varying D-FACTS settings into different BNs and focused on the power transmission lines as query nodes $\boldsymbol{Q} \subset \boldsymbol{Y}$ in the posterior $\Pr(\boldsymbol{Y} \mid \boldsymbol{e})$. The underlying concern here is potential overloading of transmission lines, which can cause sagging or drooping lines, which again can cause short circuits, blackouts, fires, and injuries.

We now discuss one such experiment in more detail, see Fig. 3. In this experiment, we use the D-FACTS settings determined by corrective SCOPF as an input into MatPower analysis. Under the simplifying assumption that we have two settings (ON versus OFF), the following computation is done twice. MatPower simulates the power flow and generates the power flow data. The CREATESTATEESTIMATOR algorithm creates a BN based on the MatPower data.

---

[9] For our probabilistic parameters we researched several datasets, all of which contained high-frequency data. The dataset that was chosen is from the Bonneville Power Authority (BPA) SCADA system, see http://transmission.bpa.gov/Business/Operations/Wind/default.aspx.

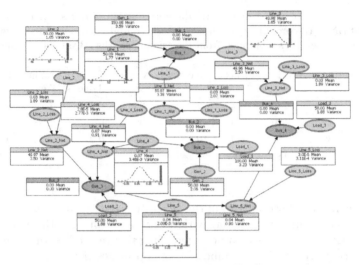

(a) D-FACTS OFF setting: The red lines denote the limit for transmission lines 4 and 5; the blue lines are limits for the other lines.
There is little probability mass beyond any of the limits, including
the two limits indicated with red lines.

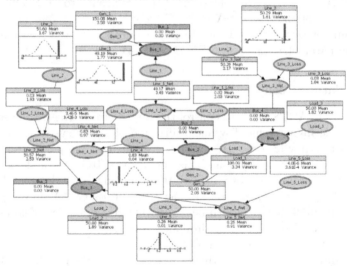

(b) D-FACTS ON setting: There is an increase in power flow for transmission lines 4 and 5. This is indicated by the significant probability
masses that go beyond the red lines, suggesting high probabilities of
overload for these lines. The other transmission lines, with blue limit
lines, are in good shape.

Fig. 3. Smart Grid BN models for a 4-bus corrective SCOPF scenario showing posterior
marginal distributions BEL(Q, e) for transmission lines $Q \in Q = \{\text{Line\_1}, \ldots, \text{Line\_5}\}$
(dashed gray curves), line limits (solid blue and red lines), and the effect of two different
D-FACTS settings (OFF in (3a) versus ON in (3b)). Figure 2 shows the structure and
node types of these BNs in more detail.(Color figure online)

The BN is then used by the STATEESTIMATION algorithm to compute posteriors BEL(Q, e) for the specific D-FACTS setting, see Fig. 3. Potential positive or negative impacts of the two D-FACTS settings (ON versus OFF) can then be identified from posteriors BEL(Q, e) (where $Q \in \boldsymbol{Q}$), and adjustments can be made in order to reduce the probability of transmission lines $\boldsymbol{Q}$ being overloaded.

Let us drill a bit deeper. The transmission line limits are shown as red and blue lines in Fig. 3a and b. The D-FACTS OFF setting, shown in Fig. 3a, is the baseline. In Fig. 3a we see that very little probability mass is beyond the limit in any posterior BEL(Q, e), for $Q \in \boldsymbol{Q}$.

Figure 3b shows the impact of D-FACTS devices ON. For transmission lines 4 and 5 there are now substantial probability masses beyond the limits (red lines) for posteriors BEL(Line_4, e) and BEL(Line_5, e). This indicates a substantial risk of overload for transmission lines 4 and 5; action may need to be taken. Further, there are in Fig. 3b (D-FACTS ON) substantial increases in the probability masses that go beyond the red lines compared to the situation in Fig. 3a (D-FACTS OFF). Thus, corrective SCOPF can be used to dispatch D-FACTS to decrease the probability of overload, and go with D-FACTS OFF in this particular scenario.

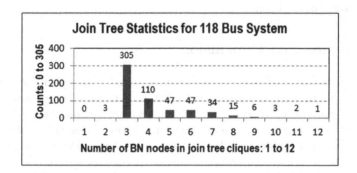

Fig. 4. BN model's JT statistics for a 118 bus system, see `case118` in Table 1, used for probabilistic state estimation. On the x-axis, the number of BN nodes (ranging from 1 to 12) in different JT cliques, or clique sizes, is shown. On the y-axis, the counts of the different clique sizes (ranging from 0 to 305) is shown.

5.3   Scalability Results

To investigate the scalability of our approach, we experiment with bus systems of varying sizes as reported in Table 1. In this table, the **Nodes** column presents the number of BN nodes; the **Convert** column presents the time it took to create the BN using CREATESTATEESTIMATOR; the **Compile** column shows the BN to JT compile time; and the **Inference** column shows the JT inference time for computing $\Pr(\boldsymbol{Y} \mid \boldsymbol{e})$. It is this last **Inference** column that is key for PSE computation time, since the two other compute time columns reflect computations that are typically performed infrequently. In addition, we emphasize that these are all CPU computation times for MATLAB implementations of the algorithms; much can be gained from using a high-performance programming language and high-performance computing hardware (see the discussion in Sect. 6).

For a more detailed understanding, we consider the BN for the 118-bus system, our largest IEEE test system. We study its JT statistics in Fig. 4. These statistics, which are greatly skewed towards small cliques and show no truly large cliques, explain why PSE using JT computation is quite fast even for the large 118-bus system. For perspective, we compare with JT experimental results reported in 2005, using 100 synthetic bipartite BNs that were irregular and had a ratio of leafs to roots of 2.4 (see [31, Table 4]). The maximal clique sizes of these 100 bipartite BNs, each with 102 binary nodes, was in the range 16–20 nodes, with an average CPU compute time of approximately 10 s [31]. While the details of these synthetic bipartite BNs and our Smart Grid BNs are clearly different, this comparison supports the realism of employing exact JT computation in the Smart Grid.

These experimental results suggest that our PSE methods can be scaled to much larger transmission systems, even when using the exact JT algorithm, as long as the topology of the power system does not change dramatically from what we have studied here.

6   Discussion and Conclusion

We now discuss key differences between traditional state estimation and our probabilistic state estimation approach. Traditional state estimation is deterministic and usually based on weighted least square optimization [1]. It takes into account uncertainties and inaccuracies in the measurements and is carried out regularly, with intervals of lengths ranging from a few seconds to a few minutes.

Exploiting recent technology advances, our results support the use of probabilistic state estimation. Key features of our probabilistic state estimation method are the following:

- It is exact, not a Monte Carlo simulation or other inexact method [2,42]. Monte Carlo simulation raises the question of how close the approximation is to the exact posterior distribution $\Pr(Y \mid e)$, and is typically hard for practitioners to accept.
- By compiling a Bayesian network to a secondary data structure (a junction tree), two iterations are needed to compute exact marginals or MPEs based on $\Pr(Y \mid e)$ [3,13,23,26,45]. This is in contrast to other state estimation algorithms, in which the number of iterations is not known ahead of time.
- It can perform multiple-hypothesis diagnosis, meaning that multiple faults (if present) can be diagnosed [29,39]. This is in contrast to other state estimation techniques, which either do not consider fault diagnosis as part of state estimation or simplifies by performing single-hypothesis diagnosis.

– It is data-driven, meaning there are machine learning algorithms that can learn from data, both from complete and incomplete datasets [4,12,16,25,37, 41,48]. Further, since the BN nodes and its structure are typically meaningful to humans, there is opportunity to combine data- and knowledge-driven modeling.
– It is GPU- and FPGA-friendly in the sense that substantially faster versions of the junction tree propagation algorithm than summarized in Table 1 exist for the GPU and FPGA platforms [44,52–54].

Compared to deterministic state estimation, there are several benefits of the probabilistic approach, such as the fact that a range of probable states that the Smart Grid may be in is computed. This includes states for which the determined settings of the D-FACTS devices might have negative or positive impacts on the transmission or distribution system. This point, along with the scalability of our approach, were clearly supported by our experimental results.

Acknowledgment. This material is based, in part, upon work supported by ARPA-E. The collaboration with Prof. Gabriela Hug and Javad Mohammadi on the integration of corrective SCOPF and PSE is also acknowledged.

References

1. Abur, A., Exposito, A.G.: Power System State Estimation: Theory and Implementation, vol. 24. CRC Press, Boca Raton (2004)
2. Aien, M., Rashidinejad, M., Kouhi, S., Fotuhi-Firuzabad, M., Najafi Ravadanegh, S.: Real time probabilistic power system state estimation. Int. J. Electr. Power Energy Syst. **62**, 383–390 (2014)
3. Andersen, S.K., Olesen, K.G., Jensen, F.V., Jensen, F.: HUGIN–a shell for building Bayesian belief universes for expert systems. In: Proceedings of the Eleventh International Joint Conference on Artificial Intelligence (IJCAI 1989), Detroit, MI, pp. 1080–1085 (1989)
4. Basak, A., Brinster, I., Ma, X., Mengshoel, O.J.: Accelerating Bayesian network parameter learning using Hadoop and MapReduce. In: Proceedings of the BigMine 2012, Beijing, China (2012)
5. Bills, G.W.: On-line stability analysis study. Technical report RP 901-1. Electric Power Research Institute (1970)
6. Borkowska, B.: Probabilistic load flow. IEEE Trans. Power Appar. Syst. **PAS–93**(3), 752–759 (1974)
7. Chavira, M., Darwiche, A.: Compiling Bayesian networks using variable elimination. In: Proceedings of the Twentieth International Joint Conference on Artificial Intelligence (IJCAI 2007), Hyderabad, India, pp. 2443–2449 (2007)
8. Chien, C.F., Chen, S.L., Lin, Y.S.: Using Bayesian network for fault location on distribution feeder. IEEE Trans. Power Deliv. **17**, 785–793 (2002)
9. Cooper, F.G.: The computational complexity of probabilistic inference using Bayesian belief networks. Artif. Intell. **42**, 393–405 (1990)
10. Darwiche, A.: Recursive conditioning. Artif. Intell. **126**(1–2), 5–41 (2001)
11. Darwiche, A.: A differential approach to inference in Bayesian networks. J. ACM **50**(3), 280–305 (2003)

12. Darwiche, A.: Modeling and Reasoning with Bayesian Networks. Cambridge University Press, Cambridge (2009)
13. Dawid, A.P.: Applications of a general propagation algorithm for probabilistic expert systems. Stat. Comput. **2**, 25–36 (1992)
14. Dechter, R.: Bucket elimination: a unifying framework for reasoning. Artif. Intell. **113**(1–2), 41–85 (1999)
15. Divan, D., Johal, H.: Distributed FACTS - a new concept for realizing grid power flow control. IEEE Trans. Power Electr. **22**(6), 2253–2260 (2007)
16. Heckerman, D., Geiger, D., Chickering, D.: Learning Bayesian networks: the combination of knowledge and statistical data. Mach. Learn. **20**(3), 197–243 (1995)
17. Holmström, K.: TOMLAB - an environment for solving optimization problems in MATLAB. In: Proceedings of the Nordic MATLAB Conference, Stockholm, Sweden (1997)
18. Hu, Y., Kuh, A., Kavcic, A., Nakafuji, D.: Real-time state estimation on microgrids. In: Proceedings of the 2011 International Joint Conference on Neural Networks, San Jose, CA, pp. 1378–1385 (2011)
19. Hu, Y., Kuh, A., Yang, T., Kavcic, A.: A belief propagation based power distribution system state estimator. IEEE Comput. Intell. Mag. **6**(3), 36–46 (2011)
20. Hug, G.: Generation cost and system risk trade-off with corrective power flow control. In: Proceedings of the 50th Annual Allerton Conference on Communication, Control, and Computing, Allerton, IL, pp. 1324–1331 (2012)
21. Hutter, F., Hoos, H.H., Stützle, T.: Efficient stochastic local search for MPE solving. In: Proceedings of the Nineteenth International Joint Conference on Artificial Intelligence (IJCAI 2005), Edinburgh, Scotland, pp. 169–174 (2005)
22. Jaakkola, T.S., Jordan, M.I.: Variational probabilistic inference and the QMR-DT database. J. Artif. Intell. Res. **10**, 291–322 (1999)
23. Jensen, F.V., Lauritzen, S.L., Olesen, K.G.: Bayesian updating in causal probabilistic networks by local computations. SIAM J. Comput. **4**, 269–282 (1990)
24. Kask, K., Dechter, R.: Stochastic local search for Bayesian networks. In: Proceedings of the Seventh International Workshop on Artificial Intelligence and Statistics (AISTATS 1999), Fort Lauderdale, FL (1999)
25. Koller, D., Friedman, N.: Probabilistic Graphical Methods: Principles and Techniques. MIT Press, Cambridge (2009)
26. Lauritzen, S., Spiegelhalter, D.J.: Local computations with probabilities on graphical structures and their application to expert systems (with discussion). J. Roy. Stat. Soc. Ser. B **50**(2), 157–224 (1988)
27. Lerner, U., Parr, R., Koller, D., Biswas, G.: Bayesian fault detection and diagnosis in dynamic systems. In: Proceedings of the Seventeenth National Conference on Artificial Intelligence (AAAI 2000), pp. 531–537 (2000)
28. Mengshoel, O.J.: Understanding the scalability of Bayesian network inference using clique tree growth curves. Artif. Intell. **174**, 984–1006 (2010)
29. Mengshoel, O.J., Chavira, M., Cascio, K., Poll, S., Darwiche, A., Uckun, S.: Probabilistic model-based diagnosis: an electrical power system case study. IEEE Trans. Syst. Man Cybern. Part A: Syst. Hum. **40**(5), 874–885 (2010)
30. Mengshoel, O.J., Darwiche, A., Uckun, S.: Sensor validation using Bayesian networks. In: Proceedings of the 9th International Symposium on Artificial Intelligence, Robotics, and Automation in Space (iSAIRAS 2008) (2008)
31. Mengshoel, O.J., Wilkins, D.C., Roth, D.: Controlled generation of hard and easy Bayesian networks: impact on maximal clique size in tree clustering. Artif. Intell. **170**(16–17), 1137–1174 (2006)

32. Mengshoel, O.J.: Understanding the role of noise in stochastic local search: analysis and experiments. Artif. Intell. **172**(8), 955–990 (2008)
33. Mohammadi, J., Hug, G., Kar, S.: A benders decomposition approach to corrective security constrained OPF with power flow control devices. In: Proceedings of the 2013 IEEE Power Energy Society General Meeting, pp. 1–5 (2013)
34. Mohammadi, J., Hug, G., Kar, S.: Fully distributed DC-OPF approach for power flow control. In: 2015 IEEE Power Energy Society General Meeting, pp. 1–5 (2015)
35. Murphy, K.P.: The Bayes net toolbox for MATLAB. Comput. Sci. Stat. **33**, 2001 (2001)
36. Park, J.D., Darwiche, A.: Complexity results and approximation strategies for MAP explanations. JAIR **21**, 101–133 (2004)
37. Pearl, J.: Probabilistic Reasoning in Intelligent Systems: Networks of Plausible Inference. Morgan Kaufmann, San Mateo (1988)
38. Poll, S., et al.: Advanced diagnostics and prognostics testbed. In: Proceedings of the 18th International Workshop on Principles of Diagnosis (DX-07), Nashville, TN, pp. 178–185 (2007)
39. Ricks, B., Mengshoel, O.J.: Diagnosis for uncertain, dynamic and hybrid domains using Bayesian networks and arithmetic circuits. Int. J. Approx. Reason. **55**(5), 1207–1234 (2014)
40. Roth, D.: On the hardness of approximate reasoning. Artif. Intell. **82**, 273–302 (1996)
41. Saluja, A., Sundararajan, P., Mengshoel, O.J.: Age-layered expectation maximization for parameter learning in Bayesian networks. In: Proceedings of the Fifteenth International Conference on Artificial Intelligence and Statistics (AISTATS 2012), La Palma, Spain, pp. 424–435 (2012)
42. Schenato, L., Barchi, G., Macii, D., Arghandeh, R., Poolla, K., Von Meier, A.: Bayesian linear state estimation using smart meters and PMUs measurements in distribution grids. In: Proceedings of the 2014 IEEE International Conference on Smart Grid Communications, pp. 572–577 (2014)
43. Schumann, J., et al.: Software health management with Bayesian networks. Innov. Syst. Softw. Eng. **9**(4), 271–292 (2013)
44. Schumann, J., Rozier, K.Y., Reinbacher, T., Mengshoel, O.J., Mbaya, T., Ippolito, C.: Towards real-time, on-board, hardware-supported sensor and software health management for unmanned aerial systems. Int. J. Progn. Health Manag. **6**(1), 1–27 (2015)
45. Shenoy, P.P.: A valuation-based language for expert systems. Int. J. Approx. Reason. **5**(3), 383–411 (1989)
46. Shimony, E.: Finding MAPs for belief networks is NP-hard. Artif. Intell. **68**, 399–410 (1994)
47. Soliman, W.M., Bahaa El Din, H.S., Wahab, M.A., Mansour, M.: Bayesian networks for fault diagnosis of large power generating stations. In: Proceedings of the 14th International Middle East Power Systems Conference, Cairo, Egypt, pp. 454–459 (2005)
48. Sundararajan, P.K., Mengshoel, O.J.: A genetic algorithm for learning parameters in Bayesian networks using expectation maximization. In: Proceedings of the Eighth International Conference on Probabilistic Graphical Models (PGM 2016), Lugano, Switzerland, pp. 511–522 (2016)
49. Weng, Y., Negi, R., Ilic, M.D.: Probabilistic joint state estimation for operational planning. IEEE Trans. Smart Grid **PP**(99), 1 (2018)
50. Yongli, Z., Limin, H., Jinling, L.: Bayesian network-based approach for power system fault diagnosis. IEEE Trans. Power Deliv. **21**, 634–639 (2006)

51. Zhang, N.L., Poole, D.: Exploiting causal independence in Bayesian network inference. J. Artif. Intell. Res. **5**, 301–328 (1996)
52. Zheng, L., Mengshoel, O.J.: Exploring multiple dimensions of parallelism in junction tree message passing. In: Proceedings of the 2013 UAI Application Workshops, pp. 87–96 (2013)
53. Zheng, L., Mengshoel, O.J.: Optimizing parallel belief propagation in junction trees using regression. In: Proceedings of 19th ACM SIGKDD Conference on Knowledge Discovery and Data Mining (KDD 2013), Chicago, IL (2013)
54. Zheng, L., Mengshoel, O.J., Chong, J.: Belief propagation by message passing in junction trees: computing each message faster using GPU parallelization. In: Proceedings of the 27th Conference in Uncertainty in Artificial Intelligence (UAI 2011), Barcelona, Spain, pp. 822–830 (2011)
55. Zimmerman, R.D., Murillo-Sánchez, C.E., Thomas, R.J.: MATPOWER: steady-state operations, planning, and analysis tools for power systems research and education. IEEE Trans. Power Syst. **26**(1), 12–19 (2011)

Deep Learning for Wave Height Classification in Satellite Images for Offshore Wind Access

Ryan J. Spick[✉] and James A. Walker

The School of Engineering and Computer Science, University of Hull,
Cottingham Road, Hull, Yorkshire HU6 7RX, UK
ryan.spick@hotmail.co.uk
https://www.hull.ac.uk/Home.aspx

Abstract. Measuring wave heights has traditionally been associated with physical buoy tools that aim to measure and average multiple wave heights over a period of time. With our method, we demonstrate a process of utilizing large-scale satellite images to classify a wave height with a continuous regressive output using a corresponding input for close shore sea. We generated and trained a convolutional neural network model that achieved an average loss of 0.17 m (Fig. 8). Providing an inexpensive and scalable approach for uses in multiple sectors, with practical applications for offshore wind farms.

Keywords: Convolutional neural networks · Deep learning
Satellite images · Off-shore wind farms

1 Introduction

Wave height at sea corresponds to the height in metres of a wave, measuring from the trough (the lowest point) to the peak (the highest point). The most common form of wave height detection uses buoy data instruments. They are by far the most robust and simple to use with transmissions occurring wireless; however, suffer from time and cost related issues. Techniques have been generated for use of radar and microwaves to measure wave height [1], however very little has been investigated into the use of RGB satellite data partially due to its inability to measure through clouds. Nonetheless, we aimed to determine a method of generating, using the vast available open source satellite data, a neural network that can learn representations of satellite data.

Satellite images provide a robust and high-resolution view of our world. Large amounts of work have been completed regarding the application of Deep Learning to land satellite data such as predicting poverty [4], however very little has been investigated on the uses for offshore applicability. Using the Copernicus API we can access Sentinel-2 satellite data for any region of the planet producing a vast data set which proves to be crucial for training neural networks that can generalise well.

© Springer Nature Switzerland AG 2018
W. L. Woon et al. (Eds.): DARE 2018, LNAI 11325, pp. 83–93, 2018.
https://doi.org/10.1007/978-3-030-04303-2_6

Sentinel-2 satellite data provide multiple spectrum data [9] (13 different colour channel ranges) and range from 10 m to 100 m spatial resolution across a 100 KM^2 region. The RGB channel, however, offers a 10 m pixel resolution which is paramount for the learning of the CNN model, by preserving as much of the visual features as possible.

Supervised Deep Learning requires an input and a corresponding label for that class, the network then attempts to learn a representation of the input data usually as a discrete class. Once a network has been trained the learned representation can be used for future classification problems [7]. Combining this method with a single dense final layer gives the ability to output continuous values, in this case corresponding to the wave height.

The scope of this work is to provide an alternative way of discovering the sea state on a given day, providing higher detailed information for the sea that is within several hundred kilometres of the shoreline and away from other measuring tools. Specifically aiming at the access to offshore wind farms as the current methods are largely centred around the use of physical, and perhaps multiple, buoys.

For safe offshore wind farm access, there are varying levels of acceptable wave height ranging from 1.2–2.5 m wave height [2] with larger and more expensive vessels required to tackle the rough waves. Table 1 shows wave heights and their meanings, red highlighted values prove dangerous to smaller vessels. We made sure to classify and train data between these values, where values past the point of 4 m would be considered extremely dangerous.

Table 1.

Wave height	Characteristics
0 m	Calm (glassy)
0 to 0.1 m	Calm (rippled)
0.1 to 0.5 m	Smooth (wavelets)
0.5 to 1.25 m	Slight
1.25 to 2.5 m	Moderate
2.5 to 4 m	Rough

Currently, access to offshore wind farms rely on buoy data at wind farms or in surrounding areas. While this is deemed reliable, with the increasing size of wind farms so will the need for more buoys situated to cover the whole area. This can add scaling costs to the construction, and with new wind farms being proposed of sizes upwards of 100 KM^2, a new method could ease the need for physical devices measuring the sea height while still providing reliable measured values. Since buoy data can be accurate with between 0–20% loss [12] and there are such small differences between safe boat travel and dangerous, our developed methods need to focus heavily on the accuracy of predictions.

Buoy data can be pulled at any time given a connection to the device, something we must consider is how often satellite data will be available for analysis. If a problem occurs on a wind farm, maintenance will need to occur at a moments notice. This could provide a drawback to the satellite methods, as currently using only one source provider of satellite data some regions will be observed twice or more every 5 days. Though possibilities to scale the proposed method to multiple satellite data providers are feasible, giving more scope for time access.

As sizes of offshore wind farms increase, so does the cost of maintenance and therefore the need to access at a moments notice a wind farm. The ability to extract data of the wave height in a large area would prove extremely useful with the potential of removing buoys altogether.

2   Processing Data

The creation of a framework was necessary due to a number of steps that needed to be taken before the data could be learned by a CNN. Using Python and the SNAP [11] (Sentinel Application Platform) library for the automated downloading and processing of the data into RGB format. Further processing was then required for the removal of anomalous extremities in the image; such as cloud or land. Leaving a clean data set which would be labelled with a backlog of buoy data that was located at the centre of our training region east of the Humber Estuary (Dowsing buoy).

For the purpose of training the neural network to classify wave height, each satellite image (10,800 × 10,800 pixel resolution) should be subdivided into 244 × 244 sections leaving a maximum of 2025 sub-sectioned images per input, minus those sections removed from the data set due to cloud/land contamination. It would be infeasible to process the entirety of the large-scale satellite images. An important concept in this methodology is the sub-dividing of the satellite images, any processing through a neural network that we refer to is using multiple sections of the overall image processed individually.

2.1   Satellite Image Cleaning

Satellite images are very rarely captured with no anomalous data, be it cloud, land or even aeroplane interference. The removal of atmospheric anomalies in satellite imaging is a key part in most pre-processing steps for the use of satellite images. Most methods are not robust enough for the level of accuracy required or depend on large levels of manual interaction [8], and so a method using deep learning was devised to remove unwanted data. To construct our data set for the purpose of removing unwanted noise from the inputs, we downloaded 30 satellite images from a similar location (Hornsea wind farm). Since the input would be far too large to process alone we subdivided the input images in to smaller sections of size 244 × 244, since each input image is of size 10,980 × 10,980 pixels we could obtain 2025 subdivided images per input, leaving a substantial amount of 60,750 images to be used in this step.

For these images, we used a manual filtering approach in order to construct a labelled database of two classes, one for images we want the model to remove/throw away and one for the pure sea images that we wish to keep. A method of adaptive thresholding [5] between neighbouring pixel values was chosen, leaving similar neighbouring pixel colours white, whereas the massively different cloud and land pixels will be set to black. Using this template of the overall input satellite image, we could pull the subdivided images into their corresponding classes.

If a satellite image was particularly noisy with cloud contamination, the thresholding variable should be high, whereas if there is not so much contamination the variable should be low. Therefore we provided a semi-automated workflow. A display was shown with multiple outputs representing low to high thresholding values (Fig. 2), we would simply click on the image that correctly identified as much of the two separated labelled points as possible. There is a possibility for error during this step as it was open to interpretation on what parts of the image should be classified as sea or land/cloud, due to the sheer number of images we would process there wasn't much concern.

These manually split images were fed into a Deep Convolutional Neural Network [6] which learned a representation of the two labelled classes to 91% accuracy, this allowed the future use of this model instead of the manual based approach. Figure 3 shows the analysis and output of one trained image. Unwanted areas being captured in a red overlay, while the sea sections we require are represented with a blue overlay. The network itself for this part was not important as the two classes provided such a significant difference in features (while remaining a binary classification problem) that even a small non-complex network could produce impressive results.

The slight amount of false negatives that will be transferred to the learning of sea tiles that may contain some noise is not necessarily bad, because when the network is tested on unseen data the noise can contribute to a reduced over fit representation. Furthermore, we use a more aggressive approach of choosing which areas to remove, purposely reducing the data set size in order to decrease the number of sea tiles that may contain small contamination being used.

3   Learning the Wave Height

Given the previous processing steps, we are left with a collection of clean sea images of 244×244 resolution accompanied with the date and time they were taken. Figure 5 shows a collection of sea images from different days - notice it's fairly easy to distinguish between the rough "wavey" images against the fairly smooth "flat" images. This is an important step for the practice of using visionary deep learning, too complex patterns can result in low levels of understanding, and thus accuracy, from the network.

To label our data, which is required by the supervised learning network, we mapped Dowsing buoy's wave height data values to the time that the satellite image was taken. This buoy resided centrally in our training dataset location.

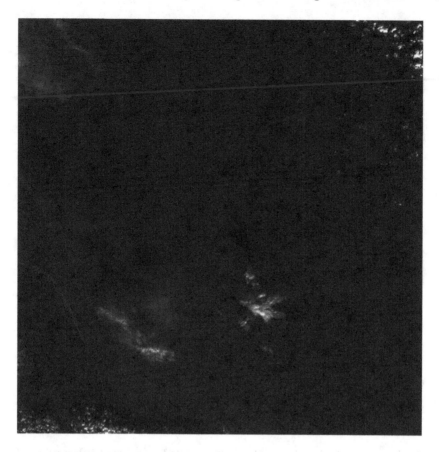

Fig. 1. Original "benchmark" satellite image to compare with outputs of the methods discussed in this section.

Each data point was rounded to 1 decimal place, which grouped the data set into a more concise collection, producing 17 data points of varying wave height, ranging from 0.5 m to 2.9 m. These values were in the range needed for sage travel to offshore wind farms, with anything greater than 3 m usually providing too much of a risk as shown in Table 1.

Deep convolutional neural networks [6] usually classify images into a discrete output, however since a continuous output is required we simply remove the final activation layer (softmax) and replace it with a dense layer with 1 output [3]. This creates a model that outputs the learned value instead of the probability of a "class" it is associated with.

The architecture of the network consisted of a pre-trained VGG16 [10] with two additional dense layers replacing the final output layer. The network was then fine-tuned [13] on the data set, while the VGG16 section was frozen. Figure 6 shows a visual representation of the network that was chosen.

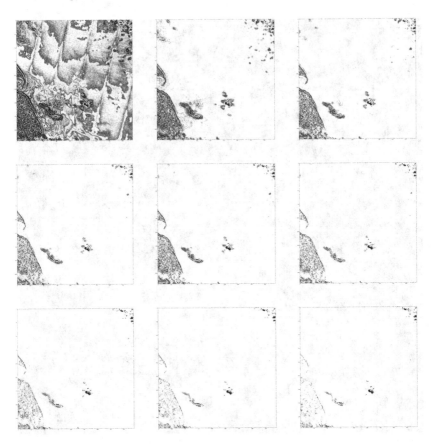

Fig. 2. Manual segregation showing 9 varying results for one input image using adaptive thresholding for distinguishing between contamination(cloud/land) and sea pixels. (Image 2 was chosen to best match the cloud and land of the original Fig.1)

While training the data we experimented with various other network models and architectures, but found the VGG16 network, while utilising fine-tuning, converged far quicker on an end result. Furthermore, the network architecture proved extremely simple to integrate with our current data set image sizes, which are very similar in size to the data set VGG16 was training on (ImageNet). We experimented with training our own model architectures from scratch consisting of around 6 layers. However, because the data set was so large, it meant that the training times proved too long for the processing power of two GTX 1080s.

The 15K preprocessed sea state images were then fed with their corresponding label outputs into the deep network (Fig. 6). Over a course of 20 epochs, we saw a gradual reduction in loss to a plateau at 0.07 m loss for the training set.

Fig. 3. RGB input image, of original Fig. 1, that has been classified using our neural network for each 244×244 sub-section. *Red* representing anomolous data and *blue* representing pure sea data. (Color figure online)

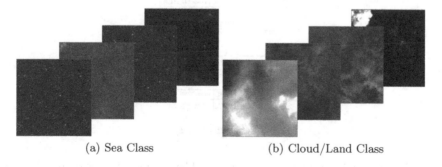

(a) Sea Class   (b) Cloud/Land Class

Fig. 4. Example of a set of image tiles from the two classes that have been separated via our convolutional neural network

Fig. 5. 20 $244 \times 244 \times 3$ random samples of varying sea state training data. Showing only slight cloud interference.

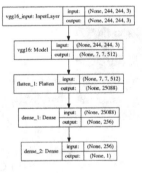

Fig. 6. Network architecture with one dense final layer for continuous output. VGG16 model is frozen.

4   Results

Initial results were run to determine if we could use a neural network to classify wave height using banded data, where we rounded each wave height into 3 classes: 1, 2 and 3 m. We attempted to train the model to classify each input into either of the three classes. Since this method provided a promising start, classifying 92% of the images into the correct three classes, we proceeded to swap out the classes for continuous outputs/regression.

The methods devised compare to some of the most accurate techniques used currently. With an average of 5% different between true and predicted values for wave height for values between 0.5 m and 3.0 m. Whereas buoy data can be seen to vary its accuracy between 0–20% [12] but cover a greater range of wave height values.

Similarly comparing the results to radar methods for wave height detection we can see that techniques currently used range around 0.21 m difference across inputs between true and predicted, so on average we perform slightly better using our technique with a gain of 0.04 m accuracy.

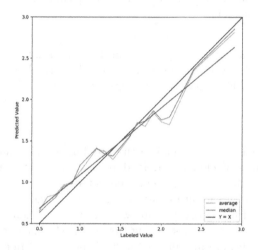

Fig. 7. Learned representation of input data, True vs Predicted labels values in metres. Plotted median and mean for comparison with blue line of best fit for both curves against black Y=X (True=Pred) perfect data representation (Color figure online)

Since a neural network is inherently difficult to understand the underlying representations we opted to plot test data to determine the accuracy of the model. Using Fig. 7 we can visualize and understand the average results of the output test set. This method of analysis was our main indicator that the network was learning over time, achieving as close to the Y=X (True = Learned) line as possible.

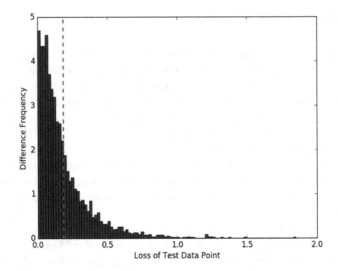

Fig. 8. Difference in classified test data between true and predicted values for 1000 data points. Average line plotted (red) for expected loss in meters for a given unknown input. (Color figure online)

Note at two points on Fig. 7, roughly at 1.4 m and 2.3 m, we experienced a large drop in loss. This was caused by a smaller than usual amount of images in these data points, with more data alleviating this issue.

5   Discussion

The techniques used proved crucial to the building of the training data, when using other methods of clouds and land removal we saw a greater disparity between clean and "dirty" images being separated.

During collection of satellite image data, it was found that a small portion of satellite data had been corrupt, this was via transmission from the satellite to the receiver. After manually scrolling through each image it was determined that the data corruption only affected a small portion of images on a specific data.

The size of the input images was also an important factor, if we lowered the image size, and subsequently obtained more data, the convolutional neural network would struggle to find patterns in the small feature data. On the opposite side, with a larger input image, we struggled to both train the network and generalise due to a smaller data set.

There are certain considerations in that since we are using RGB images, large cloud coverage proves to be an impassable problem. It was noticed that in the test set anywhere over 80% cloud coverage was simply too much to generalise on given the small amount of data remaining, which presents one large drawback of this method. From the data collected, the level of cloud coverage in a satellite image exceeding this limit was uncommon.

6   Conclusion

There has been presented a rigorous explanation of the processes and methods involved in the creation of a novel framework with the ability to consistently classify continuous (regressive) values representing the wave height in given RGB colour satellite regions.

By using multiple convolutional neural networks we automated steps that previously had manual involvement. The first network was trained to distinguish the difference between cloud contamination and land in a satellite image to a 90% level of accuracy. We created a method to train these images through a pretrained continuous convolutional neural network model which learned to an average of 0.17 m difference between actual wave height and predicted wave height.

References

1. Carrasco, R., Horstmann, J., Seemann, J.: Significant wave height measured by coherent x-band radar. IEEE Trans. Geosci. Remote Sens. **55**(9), 5355–5365 (2017). https://doi.org/10.1109/TGRS.2017.2706067
2. Stumpf, H.P., Hu, B.: Offshore wind access 2018. https://www.ecn.nl/publications/PdfFetch.aspx?nr=ECN-E-17-071. Accessed 19 Apr 2018
3. Hara, K., Vemulapalli, R., Chellappa, R.: Designing deep convolutional neural networks for continuous object orientation estimation. CoRR abs/1702.01499 (2017). http://arxiv.org/abs/1702.01499
4. Jean, N., Burke, M., Xie, M., Davis, W.M., Lobell, D.B., Ermon, S.: Combining satellite imagery and machine learning to predict poverty. Science **353**(6301), 790–794 (2016)
5. Jiang, X., Mojon, D.: Adaptive local thresholding by verification-based multithreshold probing with application to vessel detection in retinal images. IEEE Trans. Pattern Anal. Mach. Intell. **25**(1), 131–137 (2003)
6. Krizhevsky, A., Sutskever, I., Hinton, G.E.: ImageNet classification with deep convolutional neural networks. In: Advances in neural information processing systems. pp. 1097–1105 (2012)
7. LeCun, Y., Bengio, Y., Hinton, G.: Deep learning. Nature **521**(7553), 436 (2015)
8. Lin, C.H., Tsai, P.H., Lai, K.H., Chen, J.Y.: Cloud removal from multitemporal satellite images using information cloning. IEEE Trans. Geosci. Remote Sens. **51**(1), 232–241 (2013)
9. Sentinel-2: Sentinel-2 (2018). https://en.wikipedia.org/wiki/Sentinel-2. Accessed 28 July 2018
10. Simonyan, K., Zisserman, A.: Very deep convolutional networks for large-scale image recognition. arXiv preprint arXiv:1409.1556 (2014)
11. STEP: Snap. http://step.esa.int/main/toolboxes/snap/. Accessed 19 Jan 2018
12. Taylor, P.K., Kent, E.C., Yelland, M.J., Moat, B.I.: The accuracy of marine surface winds from ships and buoys. In: CLIMAR 99, WMO Workshop on Advances in Marine Climatology, pp. 59–68 (1999)
13. Yosinski, J., Clune, J., Bengio, Y., Lipson, H.: How transferable are features in deep neural networks? In: Advances in Neural Information Processing Systems, pp. 3320–3328 (2014)

Machine Learning as Surrogate to Building Performance Simulation: A Building Design Optimization Application

Sokratis Papadopoulos[1], Wei Lee Woon[2] , and Elie Azar[2($\boxtimes$)]

[1] New York University, Brooklyn, NY 11201, USA
[2] Khalifa University of Science and Technology,
Masdar Campus, PO Box 54224, Abu Dhabi, UAE
elie.azar@ku.ac.ae

Abstract. Increasing Heating, Ventilation, and Air conditioning (HVAC) efficiency is critically important as the building sector accounts for about 40% of the world's primary energy consumption. Building Performance Simulation (BPS) can be used to model the relationship between building characteristics and energy consumption and to facilitate optimization efforts. However, BPS is computationally intensive and only a limited set of building configurations can be evaluated. Machine learning techniques provide an alternative method of modeling energy consumption. While not as accurate, they can be used to perform a "first pass" evaluation of large numbers of building configurations and hence to identify promising candidates for subsequent analysis. This paper presents an initial proof-of-concept implementation of this idea. A machine learning algorithm is trained on a dataset generated using BPS, and is combined with a Genetic Algorithm (GA) based optimization to evaluate tens of thousands of building configurations in terms of energy consumption, producing designs that are very close to the optimum.

Keywords: Building design optimization · Gradient boosting
Building Performance Simulation

1 Introduction

Commercial and residential buildings account for a significant portion of the energy consumed in most developed countries [1]. Among various factors that affect building energy performance, design features such as geometry, construction materials, and choice of electro-mechanical systems, are crucial for an efficient building operation [2]. Informed decisions are therefore needed during the design phase of buildings to avoid excessive energy and carbon footprints. Building Performance Simulation (BPS) is a tool that is commonly used by designers and engineers to guide the design phase for new buildings, or the retrofitting/refurbishment of existing ones [3]. A BPS software typically takes as inputs information about a building's design and systems, operation patterns, and weather conditions. It then simulates and predicts the energy consumption levels of the modeled building over a period of time (e.g., one year) calculated at discrete intervals (e.g., daily or hourly). The mentioned capabilities make BPS a

© Springer Nature Switzerland AG 2018
W. L. Woon et al. (Eds.): DARE 2018, LNAI 11325, pp. 94–102, 2018.
https://doi.org/10.1007/978-3-030-04303-2_7

commonly-used tool to experiment with building design features, identify energy-efficient configurations and guide the design process [3].

In recent years, BPS capabilities have been combined with optimization techniques to identify the optimal building design feature combination that minimizes the energy consumption of building systems such as lighting, equipment, and Heating, Ventilation, and Air Conditioning (HVAC) systems [4–7]. Coupling schemes are typically used for this purpose to automate the process of: (1) selecting the building parameters to test, (2) simulating them in the BPS software, (3) exporting the energy predictions of the BPS, (4) analyzing the results, and (5) repeating the process until the coupling engine converges to an optimal solution. Genetic Algorithms (GA) are particularly popular for this application as they reduce the number of interactions needed to reach the optimal solution [5–7]. The main limitation of this direct coupling approach is that BPS runs are computationally expensive. As a result, the process of finding the optimal solution can become a very lengthy task, potentially forcing researchers to settle for sub-optimal solutions [8, 9].

Acknowledging this limitation, researchers have proposed replacing the BPS component with a surrogate model that mimics the predictions capabilities of a BPS model. Once a model is trained and validated, it becomes extremely efficient to run a large number of iterations through the optimization algorithm and rapidly converge to an optimal solution [8–11]. For instance, Magnier and Haghighat [11] developed a simulation-based Artificial Neural Network (ANN) to mimic building behavior, and then combine it with a multi-objective GA to successfully optimize building performance. Asadi et al. [10] followed a similar ANN with GA approach and applied it to optimize decisions related to the retrofitting of existing buildings. Finally, Gilan and Dilkina [9] proposed a Gaussian Process (GP) for prediction and active learning of building performance, followed by a GA approach for building design optimization. The authors found that the use of GP helps significantly reduce computational times when compared to ANN and other comparable methods such as Support Vector Regression (SVR).

The studies above confirm the potential of machine learning algorithms (e.g., ANN and SVR) to approximate energy consumption and motivate their use for the purpose of optimizing building design. In parallel to these efforts, more complex machine learning algorithms, notably Ensemble Learning techniques such as Gradient Boosted Regression Trees (GBRT), have become widely used in the machine learning community and are now gaining popularity in a variety of different application domains. Tree-based ensemble learning approaches such as GBRT combine statistical and machine learning methods, demonstrating high levels of accuracy and flexibility [12–14]. In a recent study by the authors of the current paper [15], the abilities of GBRT to mimic building performance estimation was benchmarked against other algorithms including ANN, SVM, GP, random forest, and extremely randomized trees. The authors found that GBRT outperformed the next best algorithm by an average of 14% and 65% for the prediction of heating and cooling loads, respectively. The findings motivated the need for the current work, which applies GBRT to train and validate a surrogate model of a BPS model, and then use GA optimization to determine the building design parameters that would minimize heating and cooling loads. To the authors' knowledge, it is the first time that GBRT is applied in the context of building design optimization. As shown later, the results confirm the adequacy of using GBRT in guiding building design to minimize heating and cooling loads. The current work builds on that of the authors [15], where the training and validation of the GBRT were documented. The main contributions of this paper are the optimization framework and application presented next, which were not published before.

2 Design of Numerical Experiment

2.1 Building Performance Simulation Dataset

The BPS dataset used in the current work was obtained from Tsanas and Xifara [16]. It was developed using a BPS of a building using the Ecotect software, where the effect of 8 input parameters of the model on the predicted heating and cooling loads of a building was tested. The dataset is commonly used in the literature and its choice eases the comparison of the results to those of previous studies.

The varied parameters include relative compactness, surface area, wall area, roof area, overall height, orientation, glazing area, and glazing area distribution, which are summarized in Table 1 along with their possible values.

Table 1. Description of dataset.

Input parameters	Units	Range of values
X1 - Relative compactness	Unitless	0.62–0.98
X2 - Surface area	m^2	514.5–808.5
X3 - Wall area	m^2	245–416.5
X4 - Roof area	m^2	110.3–220.5
X5 - Overall height	m	3.5–7
X6 - Orientation	Unitless	2–5
X7 - Glazing area	Unitless	0–0.4
X8 - Glazing distribution	Unitless	1–5[a]
Y1 - Heating load (Predicted by BPS model)	kWh/m^2	6–43.1
Y2 - Cooling load (Predicted by BPS model)	kWh/m^2	10.9–48

Note: [a]glazing configuration values are as follows: (1) 25% glazing per side, (2) 55% glazing on the north façade and 15% on the others, (3) 55% glazing on the east façade and 15% on the others, (4) 55% glazing on the south façade and 15% on the others, (5) 55% glazing on the west façade and 15% on the others.

A total of 768 combinations of input parameters were tested. For each combination, the BPS model of the building was run and the heating and cooling loads were estimated. It is important to note that all building parameters that were not mentioned in Table 1 were kept constant for all the runs. These include building location (Athens, Greece), volume (771.75 m^3), U-value for walls (1.780 $W/m^2.K$), floors (0.860 $W/m^2.K$), roofs (0.500 $W/m^2.K$), and windows (2.260 $W/m^2.K$), sedentary activity (70 W), internal heat gains in including sensible (5 W/m^2) and latent (2 W/m^2) gains, air change rate (0.5), thermostat range (19–24 °C), and operation hours (15–20 h on weekdays and 10–20 h on weekends). Additional details can be found in [16].

2.2 Gradient Boosted Regression Trees Training and Validation

GBRT is a tree-based ensemble learning algorithm that uses an iterative approach to optimally combine a set of weak learners to form a single robust and accurate model. A set of regressors (i.e., base learners) are sequentially trained with each regressor learning from the errors of the ones that precedes it. Details on GBRT algorithms can be found in [14] but broadly, let $f(x)$ be the function to be approximated using a training set $\{(x_i, y_i)\}_{i=1}^{n}$, where x_i and y_i are the ith input and target variables. For $m = 1$ to M, a set of base learners (or regressors) $h_m(x)$ and multipliers γ_m are used to improve the approximation $f_m(x)$ from that of the previous iteration, $f_{m-1}(x)$, as shown in Eq. 1. The output of the process is $f_M(x)$, which is the best approximation of $f(x)$ with the final ensemble being a weighted average of multiple base regression trees [14].

$$f_m(x) = f_{m-1}(x) + \gamma_m h_m(x) \tag{1}$$

The training of the model is performed using 80% of the dataset, while the remaining 20% is used for testing. A 10-fold cross-validation process is then performed using 9 folds to train the model under a given set of parameters and 1 fold for validation. The procedure is repeated 10 times to have each fold used for validation once prior to averaging the scores from each iteration.

For parameter tuning, we perform a grid search to identify sensible values for the parameters of the GBRT model. A summary of the parameters is shown in Table 2 along with the ranges over which they are tested, and finally the optimal combination of values that resulted in the most robust and accurate model. It is important to highlight that the experiment was repeated 100 times and the errors were averaged.

Table 2. Parameter search space and chosen values.

Parameter	Range	Optimal value
Number of estimators	[10; 20; 30; 50; 60; 100; 150; 200; 250; 300; 400; 500; 600; 800; 1200]	600
Cost function	['least squares'; 'least absolute deviation'; 'huber'; 'quantile']	'least squares'
Learning rate	[0.01; 0.02; 0.05; 0.1; 0.2; 0.25; 0.3; 0.4; 0.5]	0.1
Minimum samples split	[2–6, 8, 10, 15]	2
Maximum tree depth	[3–10, 12, 15]	5

The predictive performance of the best model from the 10-fold cross-validation process is assessed using the testing dataset, which is the 20% of the original dataset and is unseen by the model in the training phase. The performance includes the ability of the model to predict both the heating and the cooling loads of the studied building. Three metrics are computed including the mean square error (MSE), mean absolute error (MAE), and the mean percentage error [16]. For the heating loads, the observed values are 0.005 (MSE), 0.046 (MAE), and 0.23 (MAPE), while for cooling loads, the results are 0.019 (MSE), 0.090 (MAE), and finally 0.39 (MAPE). As discussed in Papadopoulos et al. [15], the findings indicate that the GBRT is capable of predicting the heating and cooling loads with very high levels of accuracy.

2.3 Genetic Algorithm Optimization

A Genetic Algorithm (GA) was used to search the space of possible building configurations, where the objective was the minimization of the heating and cooling loads for each set of parameters. More details on GAs can be found in [17, 18], but broadly, a GA is an optimization algorithm which searches for the optima in a given cost function (or "fitness" as this is known in the GA literature) without having to calculate the gradient. This is done by:

1. Randomly generating a pool of candidate solutions ("chromosomes"), which is achieved by:
   (a) Modeling the distribution of each building variable (X1–X8) as a uniform distribution between the minimum and maximum values of each variable.
   (b) Each chromosome is generated by sampling a set of these variables from each uniform distribution.
2. For each iteration or generation of the algorithm, select the fittest individuals (i.e. solutions that minimize the objective function).
3. Copies of these individuals are made, such that the total number of individuals remains constant from one generation to the next.
4. To encourage diversity and discovery of novel solutions, surviving individuals are modified by:
   (a) Randomly perturbing the constituent parameter values. In GA terminology this is known as "mutation".
   (b) Swapping parameter values between randomly chosen pairs of individuals, known as "crossover".
5. Repeat the process from step (2) for a specified number of generations, or until a termination criterion is met.

Three different variants of this experiment were conducted, where the fitness function used was (i) the heating load only, (ii) the cooling load only, and (iii) the average of the heating and cooling loads. Each generation consisted of 2000 chromosomes, and the GA was run for 15 generations (no further changes in the loadings were observed beyond this point). For each generation, the 200 fittest individuals (top 10% of the population) were passed on to the next generation, after first performing 1000 mutations and 200 crossover operations.

3 Results and Discussion

The results of the experiments are presented in Table 3. In this table, the "Reference" scores were provided for comparison and were obtained as follows: for the first two columns (heating and cooling loads only), these were the best-case loadings over the entire dataset. To facilitate direct comparison, these are the predicted scores for the best known building configurations from the training set, and were also the best over all the predicted scores. For the third and fourth columns, the scores were the loadings obtained using the building configuration which optimized the other loading. As an example, for the heating load, the reference score is the heating load obtained with the building configuration that produced the lowest cooling load.

Table 3. Heating and cooling scores – GA optimized building parameters.

Cost function used → Case ↓	Heating load only	Cooling load only	Average load	
			Heating	Cooling
Genetic Algorithm	6.006	10.89	6.006	10.93
Reference (i.e., optimal)	6.006	10.89	6.006	10.89
Difference	0%	0%	0%	0.004%

As can be seen from these results, the GBRT based prediction technique was able to produce building configurations which were extremely close to the optimal values (as determined using simulations), but at a much lower computational effort. Note that each run of the GA evaluates 30,000 different building configurations, an effort that would be computationally infeasible using conventional methods. In addition, we see that when both heating and cooling are to be optimized (a common requirement), the GA produced a building configuration which met both requirements very well. These results are preliminary and further investigations are needed to establish the validity of these findings but at this stage, they do appear to be very promising.

An advantage of using the proposed optimization approach is that it can generate more than one solution to the problem, which in this case, translates to having a number of promising combinations of building design parameters to choose from. To shed more light on these combinations, Tables 4, 5 and 6 show the top 10 chromosomes (i.e., solutions) from the last "generation" produced by the GA for three studied cost functions (i.e., heating, cooling, and average load), respectively. Please refer to Table 1 for a description of variables X1 to X8, along with their ranges and averages.

The results of Tables 4, 5 and 6 indicate little variation between the values of the parameters in the different chromosomes. More specifically, the variation is minimal within and between the tables. This indicate that the GA was converging rapidly to close-to-optimal design configuration that minimized the heating and cooling loads, and that is for three considered optimization function (i.e., heating-load only, cooling-load, combined loads). The findings confirm that the proposed GBRT and GA approach was effective at finding the optimal combinations at relatively low computation cost.

Table 4. Last chromosomes for the heating load optimization.

Chromosome	X1	X2	X3	X4	X5	X6	X7	X8
1	0.75	678.31	251.5	200.41	3.62	4.17	0.030	0.21
2	0.75	678.31	251.5	211.98	3.62	4.17	0.030	0.21
3	0.75	675.84	248.26	212.59	5.11	3.64	0.030	0.21
4	0.75	678.31	251.5	198.8	3.79	4.17	0.030	0.31
5	0.73	677.11	246.39	200.41	3.78	4.41	0.030	0.24
6	0.74	684.61	250.97	199.71	3.62	4.17	0.040	0.24
7	0.73	678.9	251.5	206.41	3.62	4.17	0.040	0.24
8	0.73	695.06	251.5	200.41	3.61	4.21	0.040	0.11
9	0.75	682.32	245.71	216.78	3.51	4.15	0.030	0.44
10	0.73	674.38	251.5	200.41	3.72	4.17	0.030	0.31

Table 5. Last chromosomes for the cooling load optimization.

Chromosome	X1	X2	X3	X4	X5	X6	X7	X8
1	0.75	687.35	255.71	204.89	4.44	2.35	0.030	0.24
2	0.75	684.82	255.05	204.89	4.44	2.42	0.020	0.3
3	0.75	682.66	255.05	204.89	4.44	2.35	0.030	0.24
4	0.75	687.35	255.05	204.89	4.42	2.35	0.020	0.24
5	0.75	687.35	251.41	207.79	4.7	2.5	0.020	0.15
6	0.75	687.35	255.71	209.46	4.44	2.35	0.030	0.3
7	0.75	687.35	255.71	209.61	4.44	2.33	0.030	0.03
8	0.74	675.78	247.13	194.08	4.65	2.35	0.030	0.3
9	0.75	678.31	245.53	214.27	5.22	2.19	0.030	0.24
10	0.75	675.78	246.16	214.27	4.49	2.49	0.010	0.27

Table 6. Last chromosomes for the combined heating and cooling load optimization.

Chromosome	X1	X2	X3	X4	X5	X6	X7	X8
1	0.75	678.31	251.56	193.87	4.63	3.64	0.030	0.24
2	0.73	689.27	251.4	189.54	4.63	4.43	0.030	0.29
3	0.75	674.38	246.44	192.81	4.99	3.64	0.030	0.11
4	0.75	678.31	252.13	194.65	5.2	3.93	0.030	0.11
5	0.74	681.13	249.24	189.93	4.63	4.3	0.030	0.15
6	0.73	678.31	253.65	189.54	5.11	3.64	0.030	0.15
7	0.75	678.31	253.6	194.65	4.65	3.64	0.030	0.11
8	0.75	678.31	251.4	189.54	4.63	4.24	0.040	0.14
9	0.74	678.31	253.65	189.54	4.63	4.32	0.030	0.14
10	0.75	678.31	251.56	189.54	4.82	4.3	0.040	0.15

On the other hand, the relative lack of diversity in the recovered building configurations suggests that this was a relatively "easy" optimization problem, which could be partly attributed to the limitations of the benchmark obtained from Tsanas and Xifara [16]. More specifically, the BPS runs were carried out using the Ecotect software, which uses the Chartered Institution of Building Service Engineers (CIBSE) Admittance method to predict the building's cooling and heating loads. This is considered a somewhat simplified technique that smooths out nonlinearities, which could be better modeled by dynamic BPS software programs, such as EnergyPlus. Nonetheless, it was important in this study to use the dataset of Tsanas and Xifara [16] to benchmark the performance of the GBRT algorithm, an essential step prior to performing the GA application.

Future research can focus on testing GBRT on a more complex dataset developed using dynamic BPS, or using data gathered from a large number of buildings. The application of the proposed framework on data from actual buildings (as opposed to BPS) can be particularly helpful to account for uncertainty in building operation and performance, which commonly leads to a gap between energy predictions made during the design phase and actual energy consumption levels observed during operation.

4 Conclusions

This paper illustrates the application of GBRT to approximate a BPS model and identify – using a GA optimization framework – the optimal building design for minimal heating, cooling, and combined heating and cooling loads. The algorithm managed to obtain solutions that are very similar to the real optimal values (obtained through simulation) but in significantly less computation time, showing promising insights for future work on the topic.

The broader finding of this work is on the applicability of combining machine learning surrogates with heuristic optimization to optimize building design in its early stages. In future iterations of this work, different algorithms will be tested in both the surrogate modelling (e.g. deep neural networks) and the optimization (e.g. particle swarm) parts. Other directions for future study include the incorporation of additional building parameters into the optimization process, such as the size and type of the heating and cooling systems or parameters related to how occupants and facility managers are operating these systems. In all such instances, the framework presented here would allow building designers to screen a large number of configurations in a fast and efficient manner to identify promising designs, which can then be analyzed more carefully using conventional BPS approaches.

References

1. Damtoft, J.S., Lukasik, J., Herfort, D., Sorrentino, D., Gartner, E.M.: Sustainable development and climate change initiatives. Cem. Concr. Res. 38(2), 115–127 (2008)
2. ASHRAE: Advanced Energy Design Guide for Small and Medium Office Buildings. American Society of Heating Refrigerating and Air-Conditioning Engineers Inc., Atlanta (2011)

3. Crawley, D.B., Hand, J.W., Kummert, M., Griffith, B.T.: Contrasting the capabilities of building energy performance simulation programs. Build. Environ. **43**(4), 661–673 (2008)

4. Papadopoulos, S., Azar, E.: Optimizing HVAC operation in commercial buildings: a genetic algorithm multi-objective optimization framework. In: Proceedings of the 2016 Winter Simulation Conference, Washington D.C. (2016)

5. Lin, S.-H.E., Gerber, D.J.: Designing-in performance: a framework for evolutionary energy performance feedback in early stage design. Autom. Constr. **38**, 59–73 (2014)

6. Tuhus-Dubrow, D., Krarti, M.: Genetic-algorithm based approach to optimize building envelope design for residential buildings. Build. Environ. **45**(7), 1574–1581 (2010)

7. Caldas, L.: Generation of energy-efficient architecture solutions applying GENE ARCH: An evolution-based generative design system. Adv. Eng. Inform. **22**(1), 59–70 (2008)

8. Papadopoulos, S., Azar, E.: Integrating building performance simulation in agent-based modeling using regression surrogate models: a novel human-in-the-loop energy modeling approach. Energy Build. **128**, 214–223 (2016)

9. Gilan, S.S., Dilkina, B.: Sustainable building design: a challenge at the intersection of machine learning and design optimization. In: Proceedings of the Workshops at the 29th AAAI Conference on Artificial Intelligence, Austin, TX (2015)

10. Asadi, E., da Silva, M.G., Antunes, C.H., Dias, L., Glicksman, L.: Multi-objective optimization for building retrofit: A model using genetic algorithm and artificial neural network and an application. Energy Build. **81**, 444–456 (2014)

11. Magnier, L., Haghighat, F.: Multiobjective optimization of building design using TRNSYS simulations, genetic algorithm, and artificial neural network. Build. Environ. **45**(3), 739–746 (2010)

12. Brillante, L., Gaiotti, F., Lovat, L., Vincenzi, S., Giacosa, S., Torchio, F., Tomasi, D.: Investigating the use of gradient boosting machine, random forest and their ensemble to predict skin flavonoid content from berry physical–mechanical characteristics in wine grapes. Comput. Electron. Agric. **117**, 186–193 (2015)

13. Zhang, Y., Haghani, A.: A gradient boosting method to improve travel time prediction. Transp. Res. Part C Emerg. Technol. **58**, 308–324 (2015)

14. Friedman, J.H.: Greedy function approximation: a gradient boosting machine. Ann. Stat. **29**, 1189–1232 (2001)

15. Papadopoulos, S., Azar, E., Woon, W.L., Kontokosta, C.E.: Evaluation of tree-based ensemble learning algorithms for building energy performance estimation. J. Build. Perform. Simul. **11**(3), 322–332 (2018)

16. Tsanas, A., Xifara, A.: Accurate quantitative estimation of energy performance of residential buildings using statistical machine learning tools. Energy Build. **49**, 560–567 (2012)

17. Goldberg, D.E.: Genetic algorithms and Walsh functions: Part I, a gentle introduction. Complex Syst. **3**(2), 129–152 (1989)

18. Goldberg, D.E.: Genetic algorithms and Walsh functions: Part II, deception and its analysis. Complex Syst. **3**(2), 153–171 (1989)

Clustering River Basins Using Time-Series Data Mining on Hydroelectric Energy Generation

Yusuf Arslan[1]([✉]), Dilek Küçük[1], Sinan Eren[1], and Aysenur Birturk[2]

[1] TÜBİTAK MRC Energy Institute, Ankara, Turkey
{yusuf.arslan,dilek.kucuk,sinan.eren}@tubitak.gov.tr
[2] Middle East Technical University, Ankara, Turkey
birturk@ceng.metu.edu.tr

Abstract. Hydropower is a significant renewable energy type with a considerable share in energy generation worldwide. As with the other common means of energy generation, hydropower is critical for the reliability and quality of electricity supply. Maintaining the reliability and quality of supply enables meeting the electricity demand of the loads adequately and efficient use of the energy resources, in addition to decreasing the related financial and environmental losses. In this paper, we target at the problem of basin clustering which is crucial for hydrological and electrical analyses regarding hydropower plants. We propose an approach based on time-series data mining on generation data of a large number of run-of-river type plants as well as of a number of representative storage type plants, in order to cluster the river basins in Turkey and present the clustering results with the related discussions. Based on these results, a new basin map is proposed which will be beneficial for enhanced hydrological and electrical analyses on hydropower and thereby for the maintenance of supply reliability and quality.

Keywords: Hydropower · Renewable energy · Time-series clustering
River basins · Data mining · Energy informatics

1 Introduction

Reliability and quality of electricity supply is a significant problem in the energy domain and it is usually considered as one of the main tasks of transmission system operators. In order to ensure the reliability and quality of the supply, different energy resources should be managed efficiently in order to (i) supply energy adequately to meet the demand, and (ii) reduce possible financial and environmental losses. Hydropower is one of the important renewable energy resources and it has a considerable share in energy generation worldwide. For instance, an

This study is carried out within the scope of the Dispatcher Information System Project (5172801) developed for TEİAŞ by TÜBİTAK MRC Energy Institute.

W. L. Woon et al. (Eds.): DARE 2018, LNAI 11325, pp. 103–115, 2018.
https://doi.org/10.1007/978-3-030-04303-2_8

average of about 20% of the energy generation in Turkey is from hydropower[1]. Hence, in order to effectively manage this resource at the country-scale, river basins should be analyzed in order to enhance the related hydrological and electrical analyses with an intention to ensure supply reliability and increase the efficiency of the related operations. Data mining methods are commonly applied to problems in several domains including energy, finance, and medicine, among others. One of these methods is time-series clustering and is among the top challenging problems in data mining research [21].

In this paper, we target at the problem of clustering river basins in Turkey in order to improve the management of hydropower. We propose a time-series clustering approach on hourly electricity generation data collected from storage type and run-of-river type hydropower plants. The aim of this paper is to propose a new basin grouping scheme, in order to facilitate the process of hydrological and security studies regarding energy. A successful grouping of the basins will be beneficial for improved analysis and planning of hydropower and hence for improved and effective maintenance of supply reliability and quality. The rest of the paper is organized as follows: In Sect. 2, background information on hydropower plants, and on basins in Turkey is presented together with the related work on hydrological classifications. In Sect. 3, the hydroelectrical energy generation dataset employed is described. The proposed clustering approach is described in Sect. 4 and experimental results with discussions are provided in Sect. 5. Finally, Sect. 6 includes conclusions of the study with future research directions.

Fig. 1. Basins of Turkey on the country map.

---

[1] https://ytbs.teias.gov.tr/.

2   Background

Hydropower is known to be a common and renewable source of energy. The corresponding hydropower plants are usually classified as: (1) storage type, (2) run-of-river (ROR) type, and (3) pumped storage. The storage-type plants have reservoirs to store excess water not utilized, while ROR type plants do not have storage capability and hence the sole objective of the latter plant type is energy generation (if sufficient streamflow is available). Besides energy generation, storage-type plants are used for the supply of drinking and irrigation water, and for flood prevention. As a further note, ROR type plants can further be classified as single-unit and multi-unit ROR type plants. Although it is not within the scope of the current study, in pumped-storage plants, excess energy is used to pump water to an elevated reservoir, so that the water can be used to generate energy using turbines during energy demand.

In energy security, it is important to characterize the production behavior region-wise for the analysis and planning of the supply. In hydroelectric energy security, river basins are usually grouped as regions. In our clustering experiments, 26 basins of Turkey are taken as a case study. The geographical visualization of the basins are shown in Fig. 1 and their names and numbers are provided in Table 1.

Table 1. Basins of Turkey and their numbers.

Name	No	Name	No
Meriç Ergene	1	Yeşilırmak	14
Marmara	2	Kızılırmak	15
Susurluk	3	Konya Kapalı (Konya Closed)	16
Ege Suları	4	Doğu Akdeniz (East Mediterranean)	17
Gediz	5	Seyhan	18
Küçük Menderes (Small Menderes)	6	Ceyhan	19
Büyük Menderes (Big Menderes)	7	Asi (Orontes)	20
Batı Akdeniz (West Mediterranean)	8	Fırat (Euphrates)	21
Orta Akdeniz (Central Mediterranean)	9	Doğu Karadeniz (East Black Sea)	22
Burdur Gölü (Burdur Lake)	10	Çoruh (Chorokhi)	23
Akarçay	11	Aras (Arax)	24
Sakarya	12	Van Gölü (Van Lake)	25
Batı Karadeniz (West Black Sea)	13	Dicle (Tigris)	26

Hydrologic classification has received a great deal of attention for a long time in various studies [4,7,8], and streamflow information was used for classification in these studies. In studies such as [2], correlation between streamflow and energy generation are demonstrated in small hydropower plants. Similarly, in [12] it is pointed out that there are similarities between hydrological and meteorological situations of the small hydropower plants in the same region. Streamflow data is used in various studies for different purposes. [16] identified basins with similar hydrologic regimes. [8] classified river regimes and identified hydrologic regions. [10] analyzed river regimes and described basin characteristics. [14] found similarities between streamflow and climatic variables. Hydrologic regionalization is studies in [11]. Streamflow clustering has drawn research attention for a long time using different methods. In [9] hierarchical clustering is employed while in [3] both hierarchical and non-hierarchical clustering are implemented. In [5], non-hierarchical (k-means) clustering is applied.

3   Hydroelectrical Energy Generation Dataset

Turkey is ranked as the seventh country among 26 European countries based on their installed hydropower capacities [20], and currently, there are 530 hydropower plants in use in Turkey. 87 of them are storage-type hydropower plants while 443 of them are ROR type hydropower plants. As pointed out in Sect. 1, these plants generate about 20% of the energy production on a yearly basis. In our previous study [1], we used a streamflow dataset which contains measurement data of 26 (representative) storage-type hydropower plants from 14 basins. All of the measurement data is collected inside the plants. The dataset used in the current study consists of two parts: (1) energy generation of storage-type hydropower plants and (2) energy generation of ROR type hydropower plants.

The first part of the dataset contains the hourly energy generation data of 21 power plants (of the 26 plants used in [1]) for two years, obtained from Dispatcher Information System [6] of TEİAŞ, where TEİAŞ is the transmission system operator of Turkey. This data of the storage-type plants is used to identify a possible correlation between streamflow and energy generation. However, it should be noted that the energy generation scheme in storage-type hydropower plants depends both on the streamflow and on the operator decisions to optimize the use of the reservoir water. Our intention is to observe clustering results based on generation data this time and identify a possible connection between streamflow and generation for this plant type. However, it is important to note that storage type hydropower plants generated electricity sometimes based on streamflow and sometimes based on energy requirements by using reservoir water. Therefore, it does not always correlate with streamflow.

For the second part of the dataset, generation data of 311 ROR type hydropower plants for two years is considered. Plants of this type do not have a storage capacity and hence the generated electricity is based on streamflow. Therefore, correlation between the electricity generation and streamflow is expected for these plants. In order to alleviate the effects of missing data in the dataset, ROR type hydropower plants with at least 16,944 corresponding records (of 706 days) and with less than 10 missing values are considered in our experiments. The missing records are filled by Last Observation Carried Forward (LOCF) method, and the continuity of the data trend is ensured. After this data processing, 36 single-unit and 164 multi-unit plants (hence, a total of 200 ROR type plants) are included in the scope of the current study.

4   Proposed Time-Series Clustering Approach

Hierarchical clustering is one of the widely used unsupervised clustering technique. Samples are grouped into a tree of clusters and distance functions are used as the clustering criteria in this method. There are several different dissimilarity measurement methods in the literature for hierarchical clustering of time series [15]. These are model-free, model-based, complexity-based and prediction-based approaches. In model-free approaches, the closeness of values at certain time points are measured in a regular manner. In model-based approaches, it is assumed that explicit parametric structures make the underlying models. In complexity-based approaches, similarity of the two time series is detected by use of shared information. In prediction-based approaches, two time series are grouped together based on the closeness of their forecasts at a certain point in the future. Besides, one of the crucial point is to choose "shape-based" or "structure-based" dissimilarity concept to calculate the distance matrix of the samples in the dataset [13]. The purpose of the shape-based dissimilarity is to make a comparison between the geometric profiles of the series while the purpose of the structure-based dissimilarity is to detect the underlying dependence structures. Shape-based hierarchical clustering is usually more suitable for short time series and, in case of long sequences with high amount of noise, structure-based dissimilarities are recommended [15]. Euclidean method is often used for shape-based clustering and correlation is used for structure-based clustering as explained with examples in [15]. In [1], we used correlation-based (structure-based) hierarchical clustering. In this paper, the same approach is employed on the hourly energy generation dataset.

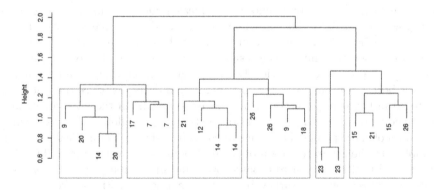

Fig. 2. Correlation-based hierarchical clustering of 21 storage type plants

5   Experimental Results and Discussions

In this section, we provide the details and results of our experiments on energy generation dataset of hydropower plants described in Sect. 3, accompanied with discussions. In the first subsection below, the results of our experiments on storage-type hydropower plants, and in the following subsection, the corresponding results on ROR type plants are presented. The experiments are all performed by using the statistical package R [18] and the basins are referred to by the numbers in Table 1.

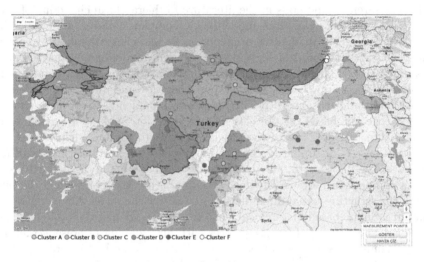

Fig. 3. Results of 21 storage type plants on the basin map of Turkey (Color figure online)

5.1   Experiments on Storage Type Hydropower Plants

In [1], we carried out experiments on 26 measurement points at storage type hydropower plants (at 14 basins) while our current dataset (see Sect. 3) contains the hourly energy generation data of 21 out of these 26 points. Hence, these 21 points were used to compare the results of this paper and that of the previous study [1]. As pointed out previously, energy generation of storage type plants depends both on streamflow data and operational decisions, since the water can be stored in the plant reservoir. Still, our experiments on these 21 points were performed to reveal possible correlations between streamflow and energy generation data of these plants. The result of correlation-based hierarchical clustering as applied to the generation data of these 21 points is shown in Fig. 2. The cluster tree in Fig. 2 is divided to 6 based on visual inspection [19] and this result is also visualized on the Turkey basin map in Fig. 3. In Fig. 3, two points in basin-23 are grouped together and some other points sharing the same basin are also clustered together. However, most of the points sharing the same basins are not clustered together as expected since the energy generation of storage-type hydropower plants does not always correlate with the streamflow.

5.2   Experiments on ROR Type HydroPower Plants

Below, we present the results of the experiments performed on ROR type plants separately based on their subclassification as single-unit and multi-unit plants. Experiments are also applied separately to all of the 200 ROR type plants.

Single-Unit ROR Type Plants. The results of clustering as applied to the energy generation data of 36 single-unit ROR type plants are presented in Fig. 6(a) in the form of a dendogram tree. The tree in Fig. 4 is divided into five clusters based on visual inspection. This result is also visualized on the basin map in Fig. 5.

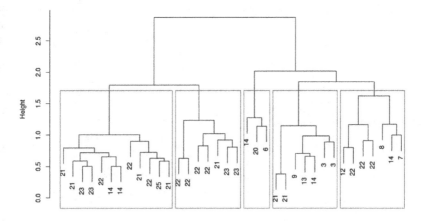

Fig. 4. Correlation-based hierarchical clustering of 36 single-unit ROR type plants

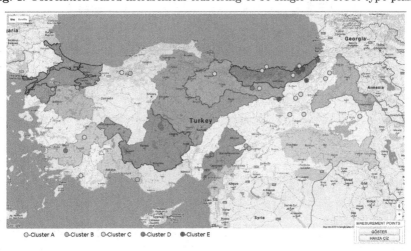

Fig. 5. Results of 36 single-unit ROR type plants on the basin map of Turkey (Color figure online)

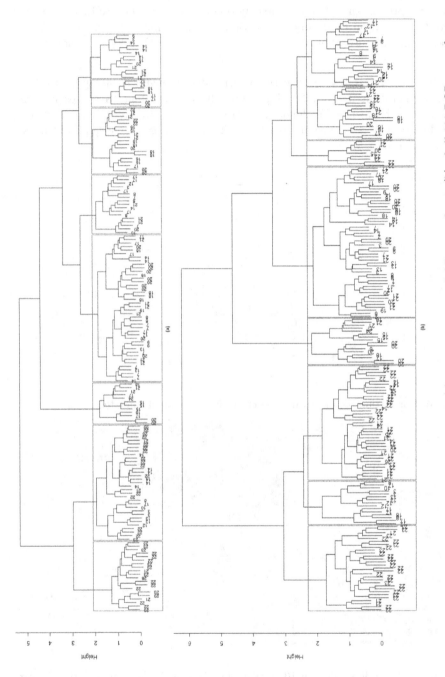

Fig. 6. Correlation-based hierarchical clustering (a) of 164 multi-unit ROR type plants, (b) of all 200 ROR type plants

All of the plants shown with purple (Cluster E) are on the north east part of Turkey and similarly all the green plants (Cluster C) are on the north east. The green plants are scattered among basins 14, 22, 23, 21 and 25 and these five basins are geographically neighbors. All the green plants in basin 21 are on the upper part of it and are close to the other plants in other basins. This observation is important since basin-21 is the largest basin in Turkey and this basin is divided into three in some studies like [17] as upper, lower, and central basins. The other plants in three other clusters (turquoise (D), pink (B) and yellow (A)) are scattered at different parts of the map.

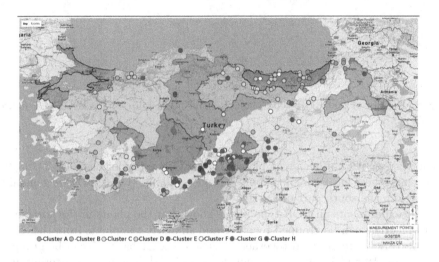

Fig. 7. Results of 164 multi-unit ROR type plants on the basin map of Turkey (Color figure online)

Multi-unit ROR Type Plants. The clustering results of 164 multi-unit ROR type plants are demonstrated as a dendogram tree in Fig. 6(a) which is divided to 8 clusters based on visual inspection. The clustering results are also visualized on the basin map in Fig. 7.

All of plants shown as yellow (Cluster A) are scattered among basins 21, 17 and 23 and these 3 basins are neighbors. The turquoise (Cluster D) plants concentrate between basins 14 and 22 and these two basins are also neighbors. All the green plants (Cluster C) are inside basins 17, 18 and 19 which are again neighbors. Purple plants (Cluster E) are generally collected in the south part of the map, although some outliers are detected in this cluster.

All ROR Type Hydropower Plants. Clustering results as applied to all of the 200 ROR type plants are shown as a dendogram tree in Fig. 6(b) which is divided to 8 based on the visual inspection. The result is also visualized on the basin map in Fig. 8. All the plants shown as yellow (Cluster A) are scattered among basins 22, 23 and in the upper part of basin 21 (upper Euphrates) which

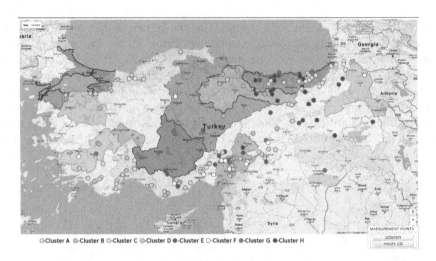

Cluster A Cluster B Cluster C Cluster D Cluster E Cluster F Cluster G Cluster H

Fig. 8. Results of 200 ROR type plants on the basin map of Turkey (Color figure online)

are neighbors. All but one green plant (Cluster C) are inside basins 14, 22, 23, 25 and 26 and these basins are neighbors. Turquoise plants (Cluster D) are generally collected in the south part of Turkey.

The experiments on ROR type plants show that the plants in the north-east part of Turkey can be grouped together and similarly the plants from the south part of Turkey are grouped together. It is shown that basin-17 can be grouped with basins 18 and 19, which was not detected in [1]. The basin clustering map proposed in [1] is also shown in Fig. 9.

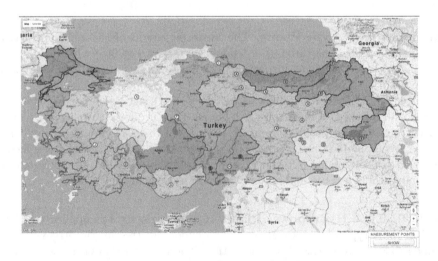

Fig. 9. Basin map proposal based on streamflow clustering presented in [1]

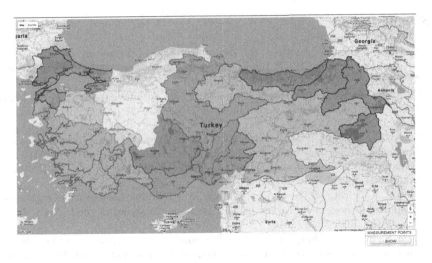

Fig. 10. New basin map proposal based on streamflow and energy generation clustering

Based on the results found in this paper and our previous work [1], a new basin grouping map is proposed as shown in Fig. 10. We envisage that, the use of this new basin grouping proposal will help facilitate related electrical and hydrological analyses to be performed on the basins in Turkey. Thereby, enhanced operation of the hydropower plants will be achieved by effective and efficient use of hydropower which is a significant renewable energy type.

6   Conclusion

Effective grouping of basins is important for the facilitation and the efficiency of the hydrological and electrical analyses and operations. As hydropower is a significant renewable energy type with a considerable share in energy generation worldwide, basin clustering will have an impact on efficient use of this energy resource. In this paper, we present the results of our correlation-based hierarchical clustering experiments on the energy generation data of hydropower plants in Turkey. The experiments are performed on genuine hourly energy generation data of two years automatically collected from both storage-type and run-of-river type hydropower plants. Based on the clustering results and on the results of previous work on streamflow data, a new basin grouping scheme is proposed. It is envisaged that this new basin grouping proposal will help facilitate and improve the related hydrological and electrical analyses and operations. As part of future work, such analyses can be conducted and the results can be verified with field data. Additionally, correlation-based hierarchical clustering can also be applied to the generation data of other power plant types. This clustering method can be applied to characterize loads of 750 substations of Turkey as residential, commercial and industrial.

References

1. Arslan, Y., Birturk, A., Eren, S.: Basin clustering of Turkey by use of monthly stream-flow data. In: IEEE International Conference on Machine Learning and Applications (ICMLA), pp. 1169–1174 (2015)
2. Basso, S., Botter, G.: Streamflow variability and optimal capacity of run-of-river hydropower plants. Water Resour. Res. **48**(10), W10527 (2012). https://doi.org/10.1029/2012WR012017
3. Bower, D., Hannah, D.M., McGregor, G.R.: Techniques for assessing the climatic sensitivity of river flow regimes. Hydrol. Process. **18**(13), 2515–2543 (2004)
4. Demirel, M.C.: Cluster analysis of streamflow data over Turkey. Ph.D. thesis, İstanbul Technical University (2004)
5. Dikbas, F., Firat, M., Koc, A.C., Gungor, M.: Defining homogeneous regions for streamflow processes in Turkey using a k-means clustering method. Arab. J. Sci. Eng. **38**(6), 1313–1319 (2013)
6. Eren, S., et al.: A ubiquitous web-based dispatcher information system for effective monitoring and analysis of the electricity transmission grid. Int. J. Electr. Power Energy Syst. **86**, 93–103 (2017)
7. Gottschalk, L., Jensen, J.L., Lundquist, D., Solantie, R., Tollan, A.: Hydrologic regions in the Nordic countries. Hydrol. Res. **10**(5), 273–286 (1979)
8. Gubareva, T.: Classification of river basins and hydrological regionalization (as exemplified by Japan). Geogr. Nat. Resour. **33**(1), 74–82 (2012)
9. Isik, S., Turan, A., Dogan, E.: Classification of river yields in Turkey with cluster analysis. In: World Environmental and Water Resource Congress, pp. 1–7 (2006)
10. Jowett, I.G., Duncan, M.J.: Flow variability in New Zealand rivers and its relationship to in-stream habitat and biota. N. Z. J. Mar. Freshw. Res. **24**(3), 305–317 (1990)
11. Kachroo, R., Mkhandi, S., Parida, B.: Flood frequency analysis of southern Africa. Hydrol. Sci. J.: I. Delineation of homogeneous regions **45**(3), 437–447 (2000)
12. Li, G., Liu, C.X., Liao, S.L., Cheng, C.T.: Applying a correlation analysis method to long-term forecasting of power production at small hydropower plants. Water **7**(9), 4806–4820 (2015)
13. Lin, J., Li, Y.: Finding structural similarity in time series data using bag-of-patterns representation. In: Winslett, M. (ed.) SSDBM 2009. LNCS, vol. 5566, pp. 461–477. Springer, Heidelberg (2009). https://doi.org/10.1007/978-3-642-02279-1_33
14. Lins, H.F.: Streamflow variability in the United States: 1931–78. J. Clim. Appl. Meteorol. **24**(5), 463–471 (1985)
15. Montero, P., Vilar, J.A., et al.: TSclust: an R package for time series clustering. J. Stat. Softw. **62**(1), 1–43 (2014)
16. Mosley, M.: Delimitation of New Zealand hydrologic regions. J. Hydrol. **49**(1–2), 173–192 (1981)
17. Özdemir, Y., Öziş, Ü., Baran, T.: Water resources development in the Euphrates-Tigris basin. In: International Perspective on Water Resources & the Environment. Dokuz Eylül University, İzmir and American Society of Civil Engineers (2013)
18. R Core Team: R: a language and environment for statistical computing (2014). https://www.r-project.org/
19. Sander, J., Qin, X., Lu, Z., Niu, N., Kovarsky, A.: Automatic extraction of clusters from hierarchical clustering representations. In: Whang, K.-Y., Jeon, J., Shim, K., Srivastava, J. (eds.) PAKDD 2003. LNCS (LNAI), vol. 2637, pp. 75–87. Springer, Heidelberg (2003). https://doi.org/10.1007/3-540-36175-8_8

20. Schlosser, P., Denina, A.: Hydro in Europe: powering renewables. EurElectric (2011)
21. Yang, Q., Wu, X.: 10 challenging problems in data mining research. Int. J. Inf. Technol. Decis. Mak. 5(04), 597–604 (2006)

Short-Term Electricity Consumption Forecast Using Datasets of Various Granularities

Yusuf Arslan[1(✉)], Aybike Şimşek Dilbaz[2], Seyda Ertekin[2], Pinar Karagoz[2], Aysenur Birturk[2], Sinan Eren[1], and Dilek Küçük[1]

[1] TÜBİTAK MRC Energy Institute, Ankara, Turkey
{yusuf.arslan,sinan.eren,dilek.kucuk}@tubitak.gov.tr
[2] Middle East Technical University, Ankara, Turkey
{aybike,seyda,karagoz,birturk}@ceng.metu.edu.tr

Abstract. It is widely known that the generation and consumption of electricity should be balanced for secure operation and maintenance of the electricity grid. In order to help achieve this balance in the grid, the renewable energy resources such as wind and stream-flow should be forecast at high accuracies on the generation side, and similarly, electricity consumption should be forecast using a high-performance system. In this paper, we deal with short-term electricity consumption forecast in Turkey, and conduct various ANN-based experiments using real consumption data. The experiments are carried out on datasets of various scales in order to arrive at a learning system that uses, as the training dataset, a convenient subset of large quantities of field data. Thereby, the performance of system can be improved in addition to decreasing the time for the training stage, so that the resulting system can be efficiently used in operational settings. The performance evaluation results of these experiments to forecast electricity consumption in Nigde province of Turkey are presented together with the related discussions. This study provides an important baseline of findings, upon which other learning systems and training settings can be tested, improved, and compared with each other.

Keywords: Electricity consumption forecast · Load forecast
Neural nets · Data mining

1 Introduction

As electricity is a commodity for which large-scale and efficient storage is not currently feasible, the generation and consumption of the electricity in the grid should be balanced. Forecasting generation is an important operational task and a research issue for renewable energy plants such as wind, solar and hydro-power

This study is carried out within the scope of the Dispatcher Information System Project (5172801) developed for TEİAŞ by TÜBİTAK MRC Energy Institute.

© Springer Nature Switzerland AG 2018
W. L. Woon et al. (Eds.): DARE 2018, LNAI 11325, pp. 116–126, 2018.
https://doi.org/10.1007/978-3-030-04303-2_9

plants. Accordingly, several data mining and machine learning algorithms have been employed in order to forecast the power to be generated by these plants. On the other hand, forecasting electricity consumption is similarly an important research issue for optimized planning and operation of the electricity grid. Excess and unplanned generation may lead to the rise of the frequency of the electricity grid above its safe band while consumption higher than expected may lead to the decrease of grid frequency below the safe band and further decreases may lead to outages, and even blackouts.

In this paper, we deal with the second part of this power system phenomenon, namely, forecast of electricity consumption. Forecasting electricity consumption (also known as *load forecasting*) has been studied for decades using machine learning methods such as ANN and SVM. ANN is also employed in the current paper for short-term electricity consumption, but with an intention to observe effects of the size and time interval of the training datasets on the forecast performance using genuine high-resolution field data. The experiments are conducted on hourly consumption data obtained from the electricity grid of Turkey. Turkey has 1,144 active transformer substations, and electricity consumption data (with a resolution of one hour) is collected from these substations on an hourly basis[1]. It approximately amounts to 60 million data samples (in five years) which should be processed to update the ANN trained models daily, to generate short-term consumption forecasts. The aim of this study is to identify the relationship between dataset size and accuracy, and to help the researchers of the domain identify the time interval for efficient model generation. Our initial experiments are conducted to forecast the short-term electricity consumption in Nigde province of Turkey by using different time granularities and dataset sizes during the training period.

The rest of the paper is organized as follows: In Sect. 2, literature review on forecasting electricity forecast is presented. In Sect. 3, the dataset used in our experiments is described and Sect. 4 includes the details of the approach used. The results of the experiments together with the discussions of these results are provided in Sect. 5. Finally, Sect. 6 concludes the paper with a summary and future work directions.

2   Literature Review

In this section, we summarize the related previous studies on electricity consumption forecast. Importance of studies in this domain from the perspective of previous studies, the definition of some of domain terms and applied methods are sum up below.

In [9], load forecasting or demand forecasting was defined as forecast of electricity need of a particular geographical region during a specific time interval. Load forecasting was described as key operation because a limited storage of electricity is offered by batteries in current technology. It was stated that, demand forecasting may prevent excess or insufficient energy production that may mean

---

[1] https://ytbs.teias.gov.tr/.

high costs and reduction of gains for electricity power suppliers. Electricity-related time series forecasting methods were categorized as linear and non-linear in this study. Linear forecasting methods were described as methods that model behavior of time series via linear functions. Non-linear forecasting methods were described as methods that model behavior of time series via non-linear functions. Non-linear methods were classified further as global and local methods based on the requirements of functions to find the characteristics. ANN was described as one of the global methods, which did not require information about the input data distribution.

In [16], sector based electricity consumption of Turkey was predicted using ANN. The countrywide total consumption (GWh), gross domestic product (GDP), index of industrial production (IIP) and population (million) data together with percentage of sector based consumption data between 1992–2011 were used for forecast. The sector based annual forecast of this study covers 10-years period (2013–2023) in yearly resolution by using past 20-years (1992–2011) data.

In [8], they proposed Rates of Load Fluctuation (RLF) and Impact Factor of the Rate (IFR) concepts for forecasting consumption of electricity. For this, they used electricity consumption data from February 1st to 10th from 2004 to 2009. The average difference between the actual value and the predicted value was 0.026 with this dataset and method used. Finally, they propose that data collection points need to be increased for a better estimate.

In [15], SVM and Least Squares Support Vector Machines (LS-SVM) were used to estimate the 24 h electricity consumption forecast. As a dataset, Inner Mongolia energy load and weather forecasts were used to train the system between 1 January and 31 December 2008. The next day weather forecast reports were used to estimate the electricity consumption to be made. 12-h forecasts were made with LS-SVM. Thus, nonlinear relations between factors were defined.

In [14], they proposed a two-stage model with multiple regression and Autoregressive Moving Average Model (ARMA) for forecasting. They used the actual data of 1995–2010 for this estimation. They compared the estimates of the proposed model with those of commercial firms. They reported that they had better results with 0 simple error in summer.

In [10], short-term load estimates were proposed. Different forecast models were created for the days of the week and holidays. The 24 h forecast was done the previous days' temperatures and average temperatures. The 24 h forecasts were delivered at 15:00 on the previous day. Therefore, data after 15:00 could not be used for forecast. It was stated that estimates have been improved with the models created in the article.

In [5], they were used Regression, Neural Net and SVM for analysis and estimation of electricity consumption. Datasets were collected from Iran and Mazandaran region between 1991 and 2013. The dataset includes data such as population, temperature, moisture as well as electricity consumption data. For Regression, Neural Net and SVM correlation coefficient were 0.996, 0.968 and 0.976 respectively. When the annual forecast was made, the error rate with the regression model was 2.48%.

In [11], two different methods were used for forecast, namely Bayesian and ANN. Short-term and long-term estimates were done for 3, 6, 12 and 18 months. For estimates, data between January 1980 and November 2013 were used. Bayesian gave better results for short-term forecast. On the other hand, ANN had better results for long term forecast.

In [1], they were tried to estimate the electric power of Algeria in the medium and long term. Multiple Linear Regressions (MLR), Artificial Neural Network Multilayer Perceptron (MLP) and Support Vector Machines Regression (SVR) were used for this. The national electricity consumption data between 2000 and 2012 was used to train the system. The years for the estimation were divided to seasons. Different models were created for each season. Estimates were made in the same seasons months belongs to previous years and past months of the same year have been used together. It is stated that modeling according to the seasons increases the accuracy of estimation. They specified that for the estimates made using the whole year, the dataset was inadequate.

In [13], the dataset which includes 1990–2015 years data was used for estimation of electricity consumption until 2030. A model with a production function of Cobb Douglas was used for estimation. Income, population and number of households were used for this model. The increments of these side parameters for the estimates were also statistically calculated. Estimates for the years 2020, 2025 and 2030 showed that electricity consumption was going to increase. But there is no information about possible error rates.

In [6], the power consumption rate was modeled and predicted in Turkey by using ANN method. Population (millions), gross national product (GNP) (billion $), exports (billion $) and imports (billion $) were employed to model the power consumption rate of Turkey (billion $ kWh). The existing data for building the model are related to the years between 1975 and 2006 which were collected yearly. Electricity consumption, population, GNP, exports and imports were forecast between 2007 and 2027 in a yearly resolution.

In [2], it was stated that forecasting models were used to set electricity generation and purchasing, establishing electricity prices, load switching, demand response and infrastructure development. Also, depending on the approach, the forecasting can be made from an aggregated level (e.g. from the electric utility side) in a top-down scheme, or from the user side, analyzing end-use activities, in a bottom-up scheme.

In [7], hourly and 24 h electricity consumption estimations were attempted using the ANN method. 24 h electric consumption data and temperature data were used to make predictions. Mean Average Percentage Error (MAPE) value for 24 h forecast was found to be 2.0%. The MAPE value for the hourly forecast was 0.78%.

In [3], the short-term electricity consumption forecasts for Turkey were made in a comparative way. ANFIS and ANN methods were used to make estimates. The data for 2009–2011 were used to make forecast. This data includes load, season and temperature information. MAPE for ANFIS was found to be 1.85. For ANN, the MAPE value was found to be 2.02.

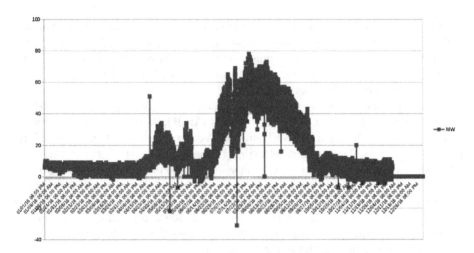

Fig. 1. Time-series of electricity consumption

Table 1. Details of horizontal chunks

Period	Coverage
10 days	21.05.2018–30.05.2018
20 days	11.05.2018–30.05.2018
1 month	01.05.2018–30.05.2018
3 months	01.03.2018–30.05.2018
1 year	31.05.2017–30.05.2018
5 years	01.06.2013–30.05.2018

3   Electricity Consumption Dataset

In the experiments, we have used electricity consumption data of Nigde province over four substations for five years period beginning from 01.06.2013 to 31.05.2018 in hourly resolution. We have 43,824 h power (MW) data for four substations in our dataset. The dataset is obtained from Dispatcher Information System [4] of Turkish Electricity Transmission Corporation (TEİAŞ), where TEİAŞ is the transmission system operator of Turkey. In Fig. 1, electricity consumption of time-series of one substation for one year period can be seen.

Abrupt peaks, sharp declines, and negative consumption (due to rare events like physical problems in transmission lines) in Fig. 1 are some of the reasons that make the consumption forecast problem very challenging. Moreover, electricity consumption can be influenced by several external factors. Special days, holidays, population size and sudden weather changes are some of the other factors that have significance on electricity consumption.

The dataset is divided into horizontal and vertical chunks for experiments. The details of the horizontal chunks can be seen in Table 1.

Chunks of 10 days, 20 days, 1 month, 3 months, 1 year and 5 years are extracted from the dataset as training data for constructing model files.

The details of vertical chunks can be seen in Table 2.

In vertical chunks, 10 days, 20 days, 1 month and 3 months data with past five years data are extracted from the dataset as training data for constructing model files.

The experiments are repeated over 3 months data and details of it can be seen in Table 3.

Table 2. Details of vertical chunks

Period	Coverage	Period	Coverage
10 days	20.05.2018–30.05.2018	20 days	10.05.2018–30.05.2018
	20.05.2017–31.05.2017		10.05.2017–31.05.2017
	20.05.2016–31.05.2016		10.05.2016–31.05.2016
	20.05.2015–31.05.2015		10.05.2015–31.05.2015
	20.05.2014–31.05.2014		10.05.2014–31.05.2014
1 month	01.05.2018–30.05.2018	3 months	01.03.2018–30.05.2018
	01.05.2017–31.05.2017		01.03.2017–31.05.2017
	01.05.2016–31.05.2016		01.03.2016–31.05.2016
	01.05.2015–31.05.2015		01.03.2015–31.05.2015
	01.05.2014–31.05.2014		01.03.2014–31.05.2014

Table 3. Details of 3 months data

Period	Coverage
1 year	01.03.2018–30.05.2018
2 years	01.03.2018–30.05.2018
	01.03.2017–30.05.2017
3 years	01.03.2018–30.05.2018
	01.03.2017–30.05.2017
	01.03.2016–30.05.2016
4 years	01.03.2018–30.05.2018
	01.03.2017–30.05.2017
	01.03.2016–30.05.2016
	01.03.2015–30.05.2015
5 years	01.03.2018–30.05.2018
	01.03.2017–30.05.2017
	01.03.2016–30.05.2016
	01.03.2015–30.05.2015
	01.03.2014–30.05.2014

In the chunks of 3 months data, past one, two, three, four and five years data for 3 months are extracted from the dataset as training data for constructing model files.

4   Methods

In this study, we have employed Artificial Neural Networks by using Neuroph library [12] in Java. ANN is relatively old but powerful method. It is especially good at nonlinear modeling. The reason that we have employed it in this study is that nonlinear characteristic of electricity consumption. We have used Multi Layer Perceptron (MLP), which is generalization of Mono Layer Perceptron, and it includes hidden layers with the difference of Mono Layer Perceptron.

The neural network is designed by implementing feed forward networks back-propagation. The network configuration of our method consists of one input, one hidden and one output layer.

Input size (N) is decided according to 24 h of one day and its repeating characteristic. Generally, electricity consumption pattern is repeated daily. Output size (N) is decided according to forecast length (24 h). The important point is to decide size of hidden layer. It is suggested to decide it empirically. We have used $2 * N + 1$ as size of hidden layer, where N is input size. The decision is based on our research on source codes of project that implemented time series forecast via neural nets. Sliding window size is set same as input size as N.

The number of iterations is set to 1000. The decision is important since big iteration size means long training time. Learning rate is set to 0.5. Maximum error is set to 0.00001. The decision is important since lower error value raises over-fitting problem. The values are decided empirically with respect to default values. Rest of the configurations is left as default values in Neuroph library.

Two error measurements are employed to compare the results, namely, Mean Absolute Error (MAE) and Normalized Mean Absolute Error (NMAE). MAE, which is the average of absolute differences between actual and predicted values, is calculated as given in Eq. 1.

$$MAE = \frac{1}{n} \sum_{j=1}^{n} \mid y_j^a - y_j^p \mid \tag{1}$$

where y^a denotes actual values and y^p denotes forecast values.

NMAE is calculated as given in Eq. 2.

$$NMAE = \frac{\sum_{j=1}^{n} \mid y_j^a - y_j^p \mid}{\mid \sum_{j=1}^{n} y_j^a \mid} * 100 \tag{2}$$

where y^a denotes actual values and y^p denotes forecast values. NMAE normalizes MAE by using actual values.

5   Experimental Results and Discussions

As described in Sect. 3, in the experiments, the electricity consumption data of four substations in Nigde province is exploited. The data is a collection of five years with hourly resolution.

As a preprocessing, the consumption values in dataset are normalized in [0–1] range with classical normalization method by using max and min values. For test case, one day data in hourly resolution is used.

When forecasting one-day ahead, different time granularities are used for model generation. One-day ahead forecast is implemented by using constant window. Last 24 h actual consumption data is supplied as input to the network and forecast of next 24 h is acquired as output. One day (31.05.2018) in hourly resolution is used to generate test network and the next day (01.06.2018) in hourly resolution is predicted.

In our experiments, the dataset is divided as horizontal and vertical chunks. In horizontal partition, the dataset is divided 10 days, 20 days, 1 month, 3 months, 1 year and 5 years chunks. In vertical partition, the dataset is divided 10 days, 20 days, 1 month and 3 months chunks that contain past 5 years data in each year for the specified chunks. The dataset of 3 months data consists of past 5 years data in each year for the specified chunks.

NMAE measurements are calculated in YTBS daily. A screen-shot of forecast module of YTBS can be seen in Fig. 2.

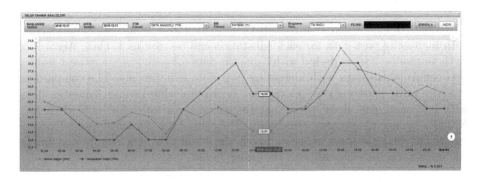

Fig. 2. Forecast values and error rates in YTBS for one substation

Error rates of horizontal chunks for each substations and averages of periods can be seen in Table 4. According to error rates, the lowest error rates are obtained for 1 year and 1 month period in horizontal.

Results of experiments on models that are generated by use of vertical chunks for each substation can be seen in Table 5. According to error rates, the lowest error rates are obtained for 3 months period in vertical. The experiments are repeated to understand the effect of vertical partition with increasing time interval.

Table 4. Error rates of horizontal chunks for substations. In this table and the other tables henceforth, in the first column, P stands for Period, d stands for days, m stands for months, and y stands for years.

P	Substation-1		Substation-2		Substation-3		Substation-4		Averages	
	MAE	NMAE	MAE	NMAE	MAE	NMAE	MAE	NMAE	MAE	NMAE
10 d	**5.04**	**27.44**	11.03	259.44	2.33	5.28	4.73	17.86	5.78	77.50
20 d	5.37	29.26	**2.80**	119.32	3.17	7.16	7.27	27.43	4.65	45.79
1 m	11.11	60.49	2.95	**69.38**	3.35	7.58	4.26	16.09	5.41	**38.38**
3 m	6.20	33.78	5.55	130.66	**2.03**	4.60	5.52	17.09	4.82	46.53
1 y	8.10	44.11	4.32	101.72	2.35	5.30	**2.92**	**11.05**	**4.42**	40.54
5 y	10.90	59.35	7.27	171.22	1.93	**4.38**	3.04	11.49	5.78	61.61

Table 5. Error rates of vertical chunks for substations

P (5 y)	Substation-1		Substation-2		Substation-3		Substation-4		Averages	
	MAE	NMAE	MAE	NMAE	MAE	NMAE	MAE	NMAE	MAE	NMAE
10 d	9.96	54.23	2.99	70.43	2.25	5.09	4.30	16.24	4.87	36.50
20 d	9.85	53.64	4.70	112.73	2.72	6.16	**2.98**	**11.27**	5.06	45.95
1 m	10.08	54.87	3.64	85.70	**1.94**	**4.38**	4.59	17.33	5.06	40.57
3 m	**9.17**	**49.90**	**2.95**	**69.47**	2.56	5.79	3.02	11.40	**4.42**	**34.14**

Results of experiments on models that is generated by use of 3 months data with increasing past data for each substation can be seen in Table 6. According to these results, 3 years period with 3 months data for each year gives best results between 3 months data with increasing past data experiments. The comparison of error rates between vertical chunks, horizontal chunks and 3 months data with increasing past data shows that the lowest error rates, for both MAE and NMAE, belong to 3 years period with 3 months data. 3 months period covers months of spring seasons, namely, march, april and may. The results may imply the importance of seasonality in the experiments.

Table 6. Error rates of 3 months data with increasing past data for substations

P (3 m)	Substation-1		Substation-2		Substation-3		Substation-4		Averages	
	MAE	NMAE	MAE	NMAE	MAE	NMAE	MAE	NMAE	MAE	NMAE
1 y	**6.20**	**33.78**	5.55	130.66	2.03	4.60	5.52	17.09	4.82	46.53
2 y	7.38	40.16	5.41	127.21	2.37	5.36	11.51	13.3	6.67	46.51
3 y	8.53	46.44	**2.67**	**62.88**	2.16	4.88	4.09	15.44	**4.36**	**32.41**
4 y	8.44	45.92	3.10	72.92	**1.91**	**4.33**	6.86	25.88	5.10	37.26
5 y	9.17	49.90	2.95	69.47	2.56	5.79	**3.02**	**11.40**	4.42	34.14

3 years period with 3 months data contains 6,552 data samples. Turkey has 1,144 active transformer substations with 60 million data samples as stated in Sect. 1 and use of 3 years period with 3 months data as training data for model generation means use of approximately 7.5 million data samples for model generation.

6   Conclusion

In conclusion, artificial neural networks have been implemented as non-linear method in this study. Univariate data is exploited in the experiments and results are compared by using different error measurements techniques. According to results of experiments, one year and one month periods give lowest error rates in horizontal and three months period for five years period gives lowest error rates in vertical. The comparison of one year in horizontal and three months in vertical reveals that vertical partition gives better results than horizontal partition. Further experiments have been conducted on three months data by using increasing past data. The best results are obtained by three years period with three months data. The chunk that contains three months data from past three years is delivered lowest error rates between all experiments. This chunk consists of data from months of spring season. Therefore, the results may imply the importance of seasonality.

As a future work, the experiments can be extended on various deep learning algorithms to understand the effect of sizes of datasets on accuracy of consumption forecast and to detect the best size and coverage of training data for model generation. Besides, the energy consumption is known to have seasonality and trends factors, further analysis can be conducted on the impact of these components on accuracy.

References

1. Ahmia, O., Farah, N.: Multi-model approach for electrical load forecasting. In: 2015 SAI Intelligent Systems Conference (IntelliSys), pp. 87–92. IEEE (2015)
2. Campillo, J., Wallin, F., Torstensson, D., Vassileva, I.: Energy demand model design for forecasting electricity consumption and simulating demand response scenarios in Sweden. In: 2012 4th International Conference in Applied Energy, Suzhou, China, 5–8 July 2012
3. Çevik, H.H., Çunkaş, M.: A comparative study of artificial neural network and ANFIS for short term load forecasting. In: 2014 6th International Conference on Electronics, Computers and Artificial Intelligence (ECAI), pp. 29–34. IEEE (2014)
4. Eren, S., et al.: A ubiquitous web-based dispatcher information system for effective monitoring and analysis of the electricity transmission grid. Int. J. Electr. Power Energy Syst. **86**, 93–103 (2017)
5. Karimtabar, N., Pasban, S., Alipour, S.: Analysis and predicting electricity energy consumption using data mining techniques-a case study IR Iran-Mazandaran province. In: 2015 2nd International Conference on Pattern Recognition and Image Analysis (IPRIA), pp. 1–6. IEEE (2015)

6. Kavaklioglu, K., Ceylan, H., Ozturk, H.K., Canyurt, O.E.: Modeling and prediction of Turkey's electricity consumption using artificial neural networks. Energy Convers. Manag. **50**(11), 2719–2727 (2009)
7. Kotur, D., Žarković, M.: Neural network models for electricity prices and loads short and long-term prediction. In: 2016 4th International Symposium on Environment Friendly Energies and Applications (EFEA), pp. 1–5. IEEE (2016)
8. Ma, R., Jiang, F., Song, J., Chen, H., Dong, H.: The short-term load forecasting based on the rate of load fluctuation. In: 2011 International Conference on Intelligent Computation Technology and Automation (ICICTA), vol. 1, pp. 983–986. IEEE (2011)
9. Martínez-Álvarez, F., Troncoso, A., Asencio-Cortés, G., Riquelme, J.C.: A survey on data mining techniques applied to electricity-related time series forecasting. Energies **8**(11), 13162–13193 (2015)
10. Rao, M., Soman, S., Menezes, B., Chawande, P., Dipti, P., Ghanshyam, T.: An expert system approach to short-term load forecasting for Reliance Energy Limited, Mumbai. In: 2006 IEEE Power India Conference (2006)
11. Rivero, C.R., Sauchelli, V., Patiño, H.D., Pucheta, J.A., Laboret, S.: Long-term power consumption demand prediction: a comparison of energy associated and Bayesian modeling approach. In: 2015 Latin America Congress on Computational Intelligence (LA-CCI), pp. 1–6. IEEE (2015)
12. Sevarac, Z., et al.: Neuroph-Java neural network framework (2012). Accessed 01 July 2012. http://neuroph.sourceforge.net/
13. Vu, N.H.M., Khanh, N.T.P., Cuong, V.V., Binh, P.T.T.: Forecast on Vietnam electricity consumption to 2030. In: 2017 International Conference on Electrical Engineering and Informatics (ICELTICs), pp. 72–77, October 2017. https://doi.org/10.1109/ICELTICS.2017.8253238
14. Wang, E., Galjanic, T., Johnson, R.: Short-term electric load forecasting at Southern California Edison. In: 2012 IEEE Power and Energy Society General Meeting, pp. 1–3. IEEE (2012)
15. Wu, J., Niu, D.: Short-term power load forecasting using least squares support vector machines (LS-SVM). In: Second International Workshop on Computer Science and Engineering, WCSE 2009, vol. 1, pp. 246–250. IEEE (2009)
16. Yetis, Y., Jamshidi, M.: Forecasting of Turkey's electricity consumption using artificial neural network. In: 2014 World Automation Congress (WAC), pp. 723–728. IEEE (2014)

Intelligent Monitoring of Transformer Insulation Using Convolutional Neural Networks

Wei Lee Woon[1(✉)], Zeyar Aung[1], and Ayman El-Hag[2]

[1] Department of Computer Science, Masdar Institute, Khalifa University of Science
and Technology, P.O. Box 127788, Abu Dhabi, UAE
{wei.woon,zeyar.aung}@ku.ac.ae
[2] Department of Electrical and Computer Engineering, University of Waterloo,
200 University Avenue West, Waterloo, ON N2l 3G1, Canada
ahalhaj@uwaterloo.ca

Abstract. The ability to monitor and detect potential faults in smart grid system components is extremely valuable. In this paper, we demonstrate the use of machine learning techniques for condition monitoring in power transformers. Our objective is to classify the three different types of Partial Discharge (PD), the identify of which is highly correlated with insulation failure. Measurements from Acoustic Emission (AE) sensors are used as input data. Two broad machine learning based approaches are considered - the conventional method which uses a predefined feature set (Fourier based), and deep learning where features are learned automatically from the data. The performance of deep learning compares very favorably to the traditional approach, which includes ensemble learning and support vector machines, while eliminating the need for explicit feature extraction from the input AE signals. The results are particularly encouraging as manual feature extraction is a subjective process that may require significant redesign when confronted with new operating conditions and data types. In contrast, the ability to automatically learn feature sets from the raw input data (AE signals) promises better generalization with minimal human intervention.

Keywords: Machine learning · Smart grid · Deep learning
Power transformer · Convolutional neural networks · Partial discharge

1 Introduction

1.1 Background and Motivation

Power transformers are amongst the most expensive and important assets in transmission and distribution systems, and can remain in service for over a century [1,2]. Failure in a typical 100 MVA power transformer is a very serious event and can result in severe disruptions to business activities and energy efficiency initiatives, which could have serious environmental and financial consequences.

© Springer Nature Switzerland AG 2018
W. L. Woon et al. (Eds.): DARE 2018, LNAI 11325, pp. 127–136, 2018.
https://doi.org/10.1007/978-3-030-04303-2_10

The aging of the transformer insulation system during its operational life is a natural phenomenon as transformers are constantly subjected to a range of electrical, mechanical, and thermal stresses [3]. Degradation in the insulation system frequently manifests as a localized dielectric breakdown known as a Partial Discharge (PD) [4]. Hence, detecting and assessing PDs can help prevent more serious transformer failure. Unfortunately, conventional monitoring techniques require the physical presence of a site engineer to diagnose problems. More recently, machine learning techniques [5] have been proposed as a way of automatically detecting PDs in transformer insulation, which could be significantly more cost-effective and reliable.

PDs can be detected in a number of ways. One interesting option is via the use of acoustic emission (AE) sensors [4], which are noninvasive, cost effective and easily installed by magnetically attaching the sensors to the transformer's tank wall - a procedure that can be undertaken even when the transformer is energized.

Furthermore, PDs can originate from different sources, each of which is indicative of different fault types and severities. As such, in this paper, we will focus on the *classification* of different PD types using machine learning methods to analyze data generated by AE sensors.

1.2   Novelty and Objectives

There have been prior attempts at using AE signals to identify PDs (see for e.g. [1,6]). In [7], a wide band piezoelectric transducer, DC-1 MHz, was used to measure eight simulated PD types. An artificial neural network was used to classify Power Spectral Density (PSD) and Short Time Fourier Transform (STFT) features, resulting in recognition rates of over 90%. However, measurement conditions such as the oil temperature were not considered. In more recent work [6], AE sensors were used to study the effect of increasing the tank size, the presence of barriers in the insulating medium and oil age on PD detection capability. High recognition rates (96–100%) were obtained using spectral and statistical features, even when barriers were placed between the PD source and the AE sensor. However, the recognition rate dropped to 60–88% when old insulation oil was used.

In previous studies conducted by the authors of this paper, the classification of PDs initiated under different experimental conditions were presented [5,8]. In agreement with other prior work, when the training and test datasets consisted of samples collected under similar conditions, high PD classification rates were achieved. However, recognition rates were lower when test sets collected under different experimental conditions were used.

In this study, our aim is to study the use of deep learning techniques on this challenging problem. These methods have been used with great success to analyze natural signals such as sound waves and digital images, and it would be very interesting to gauge the applicability of deep learning methods in the present context, and to characterize the performance characteristics *vis-a-vis* conventional methods.

2   Methods and Data

2.1   Classification

Classifications algorithms intelligently discriminate between objects or instances
from different classes by identifying a suitable decision boundary (or boundaries)
in the feature space in which the instances are embedded. A variety of different algorithms have been devised, with significant differences in performance
characteristics.

As benchmarks, we use the methods presented in [5], which are the Random
Forest, Gradient Boosting, SVM, Decision Tree and LDA algorithms. A detailed
explanation of each of these algorithms is beyond the scope of the paper and the
interested reader is referred to [5] as well as the many excellent references in the
literature, for e.g. [9,10].

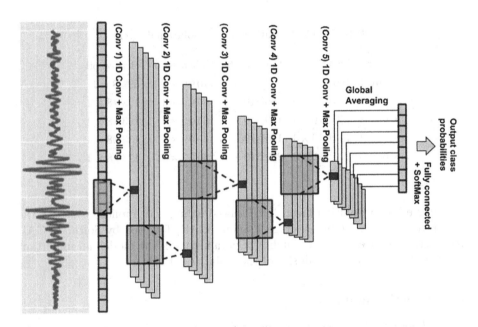

Fig. 1. 1D-CNN - Model architecture

2.2   1D-Convolution Neural Networks

A Convolutional Neural Network (CNN, or ConvNet) is a type of deep learning
architecture that is commonly used in computer vision, but which can also be
used for other forms of signal processing. CNNs are feedforward neural networks
which contain one or more convolutional layers; these are composed of specialized
units or *kernels* which filter inputs from a finite subset of the input data.

Traditional neural networks are "fully connected", i.e. all the nodes in one
layer are connected to all of the nodes in the next - this has two big disadvantages

(i) it greatly increases the number of trainable parameters and, as a consequence, the amount of training data that is required (ii) when processing signals, the presence of a feature (say, a spike in a time series) may be important but not the exact position of the feature in the time series. A fully connected network will treat such occurrences as entirely different phenomena.

CNNs do not suffer from the above shortcomings, and are able to learn customized features automatically from the data itself, as opposed to the use of manually selected features, such as Fourier coefficients. Many of the notable applications of CNNs have been in machine vision, but 1-dimensional CNNs can equally be applied to time series and sequence data. Motivated by the advantages mentioned above, this is the approach adopted in this study.

The architecture used in this study is shown in Fig. 1. The configurations for each convolutional layer were determined using a randomized parameter search, and are shown in Table 1.

Table 1. Configurations for convolutional layers

Layer	Location 4	Location 9	Unheated oil	Heated oil
Conv 1	$100@10 \times 1$	$100@10 \times 1$	$110@9 \times 1$	$89@8 \times 1$
Conv 2	$150@10 \times 1$	$150@10 \times 1$	$165@9 \times 1$	$133@8 \times 1$
Conv 3	$150@10 \times 1$	$150@10 \times 1$	$165@9 \times 1$	$133@8 \times 1$
Conv 4	$150@10 \times 1$	$150@10 \times 1$	$165@9 \times 1$	$133@8 \times 1$
Conv 5	$200@10 \times 1$	$200@10 \times 1$	$220@9 \times 1$	$178@8 \times 1$

After each convolution layer, 3×1 Max-Pooling was applied, while for the final layer we used Global Average Pooling (GAP) followed by Softmax. GAP was used to reduce the total number of parameters in the model, an important consideration given the small size of the training data sets.

2.3   Experimental Setup

Three different PD classes were generated, namely: (i) Discharges from a sharp point to ground plane (ii) Surface discharges (iii) Discharges from a void in the insulation. A $1 \times 1 \times 0.5$ m tank filled with aged oil received from a local utility company was used to conduct the experiments.

An AE sensor with a resonance frequency of 150 kHz and bandwidth of 100–450 kHz was used, while data acquisition was performed using an oscilloscope interfaced with Matlab^{TM}, with the sampling frequency set to 10M sample/sec for a window of 2500 samples (250 μs). A magnetic holder is used to attach the AE sensor to the tank wall and silicone grease was applied to improve acoustic coupling.

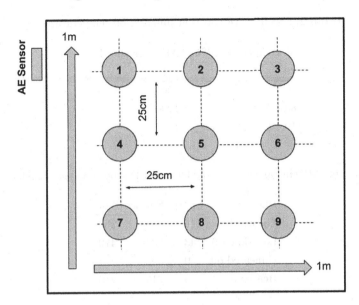

Fig. 2. Location of PDs relative to AE sensor [5]

2.4   Measurement Conditions

As mentioned previously, the more challenging classification problem is when data from the different PD types were collected under variations in the measurement conditions. In this paper we focus on two such variations:

- **Location of the PD Source.** PDs can be initiated at any physical point in the insulation system. Figure 2 shows the layout of the measurement tank annotated with the location of the AE sensor. As can be seen, the tank contains nine holes which allow the PD sources to be generated at different locations. For the results in this paper, the different PD types were simulated at locations 4 and 9.
- **Temperature of Insulating Oil.** The temperature of a power transformer's oil varies according to operating conditions, and can be as high as $80\,°C$. This can affect the classifier performance as AE waves in oil have different propagation speeds at different oil temperatures. Table 2 shows the effect of increases in the oil temperature.

Note that in [5], the presence of a barrier in the insulation medium was also studied but the corresponding data is less complete and this was excluded from this paper due to space limitations. The total number of measurements for each of the PD types and measurement conditions are shown in Table 3.

Table 2. Speed of AE waves at different oil temperatures [4]

Oil Temperature (°C)	AE wave speed (m/s)
20	1413
50	1300
80	1200
110	1100

Table 3. Count of PD instances measured for each PD type and measurement condition

PD Type	Sharp	Surface	Void
Location #4	116	253	253
Location #9	143	233	472
Unheated oil	0	393	473
Heated oil	0	216	216

2.5 Data Pre-processing and Analysis

Computational Tools. All the development and data processing in this paper was performed using the Python programming language. Python has a concise, intuitive syntax which allows rapid development of a variety of different applications and tools. It also has an extremely wide variety of toolboxes which can support most scientific and technical computing tasks. Classification algorithms were implemented using *Keras*[1] for the CNNs, and *scikit-learn*[2], a commonly used and flexible Python machine learning toolkit.

Feature Extraction. While CNNs are capable of learning customized feature sets, the benchmark classification algorithms require a more traditional feature extraction process, where instances to be classified are first converted into a feature vector which adequately characterizes these instances. As per [5], we used a Discrete Fourier Transform (DFT) based method to convert each time series into PSD histograms. The PSDs were extracted from the time series using the *periodogram* function from the well known *SciPy*[3] Python toolkit.

Validation. Two numerical metrics were used to validate the performance of the classification algorithms:

1. **Accuracy:** the number of instances correctly classified as a percentage of the total number of instances in the test set. Accuracy is a reasonable metric to

---

[1] https://keras.io/.
[2] http://scikit-learn.org.
[3] http://www.scipy.org.

use where the number of training instances in the different classes are fairly well balanced, as is the case in this study.

2. **AUC:** The **A**rea **U**nder the **C**urve of the Receiver Operating Characteristic (ROC) curve for a given classifier. The ROC curve is the plot of true *vs* false positives, and allows the quality of a classifier to be evaluated. While accuracy only evaluates the operation of the classifier at a particular decision threshold, ROC rates overall performance and can help to detect cases where a particular classifier may be highly accurate, but is also very sensitive to minor shifts in the data or parameters.

Finally, for the cases where classifiers are trained and tested using data from the same recording session, cross-validation is used in combination with these numerical scores to ensure that a fair evaluation of the performance is obtained. More detailed discussions of these evaluation techniques can be obtained from [9].

3   Results

The results produced after conducting the experiments described earlier will now be presented and discussed.

3.1   Effect of Measurement Location

The first set of experiments targeted the effect of PD source location relative to the acoustic sensors. The results of these experiments are presented in Table 4. Some general observations:

1. The classification performance for experiments involving only individual locations (as measured using cross validation) were very high. Better scores were obtained in the case of Location 9.
2. When only a single measurement location at a time was considered (cross validation), the highest accuracies were obtained using the SVM classifier and Random Forest, but the CNN classifier was always very close.
3. However, when the classifiers were trained and tested using data collected at different locations, there was a sharp drop in accuracy for all classifiers. However, the CNN classifier produced the highest accuracy by a significant margin when classifying partial discharges at Location 9, and was very close to the best classifier (Random Forest) for Location 4.
4. The pattern in the AUC scores is similar to accuracy, but CNN obtained the top score for the cross validation case with Location 4.

Table 4. Classification accuracies and AUC scores (different sensor locations). Highest scores per column are shown in bold.

Accuracy	4 (CV)	9 (CV)	4→9	9→4
CNN	88.8%	95.2%	**85.5%**	71.9%
Random Forest	85.2%	**99.6%**	80.0%	**73.2%**
Gradient Boosting	84.7%	99.1%	75.0%	71.4%
SVM	**89.3%**	97.1%	75.4%	72.1%
Decision Tree	81.4%	96.2%	70.2%	59.8%
LDA	79.6%	95.5%	72.9%	63.9%
AUC scores				
CNN	**0.94**	0.99	0.76	0.70
RandomForest	0.89	**1.00**	**0.87**	**0.83**
GradientBoosting	0.91	**1.00**	0.83	0.78
SVM	0.93	**1.00**	0.71	0.76
DecisionTree	0.82	0.95	0.62	0.51
LDA	0.81	0.97	0.58	0.61

Table 5. Classification accuracies and AUC scores (different oil temperatures). Highest scores per column are shown in bold.

Accuracy	Cold (CV)	Hot (CV)	Cold→Hot	Hot→Cold
CNN	97.7%	94.1%	**97.1%**	77.1%
RandomForest	99.2%	**99.7%**	89.8%	84.6%
GradientBoosting	**99.4%**	97.4%	92.6%	**85.7%**
SVM	97.8%	94.9%	50.0%	51.7%
DecisionTree	96.2%	97.5%	76.2%	72.6%
LDA	96.9%	94.8%	67.8%	57.5%
AUC scores				
CNN	0.99	0.97	**0.99**	0.88
RandomForest	**1.00**	**1.00**	**0.99**	**0.99**
GradientBoosting	**1.00**	**1.00**	0.96	0.95
SVM	0.99	0.99	0.42	0.70
DecisionTree	0.96	0.98	0.76	0.74
LDA	0.99	0.96	0.65	0.68

3.2 Effect of Oil Temperature

The experiments were repeated at a single fixed location, but at different oil temperatures. Results are presented in Table 5. Some observations:

1. As with the previous set of results, the cross validation results were high across all the classifiers.
2. For the cross-condition experiments, the CNN classifier had the highest score when classifying partial discharges in heated oil.
3. However, the performance on the unheated oil was a little disappointing as the Random Forest and Gradient Boosting techniques both scored significantly higher, though CNN still outperformed the methods used in [8] (SVM, Decision Tree and LDA). However this disparity is not surprising as there were far more training samples for the unheated oil (on which the classifier for the heated oil used here is trained).

4 Discussions

Overall, the results of the experiments were promising and strongly motivate the need for future studies on the use of deep learning to identify faults in power transformers.

While the results obtained using convolutional neural networks were not a drastic improvement over Random Forest and Gradient boosting they were superior in two of the four cross-condition experiments (which are more important) and were very comparable in the other cases.

Note that this is actually an impressive result as there was unfortunately only very little data available, whereas CNNs typically require much larger training sets. In contrast, ensemble classifiers are considered top of the range amongst the "conventional" machine learning techniques and would be expected to produce very good performance in such circumstances. This view is substantiated by the observation that CNNs performed best in the cases where more data was available.

Finally, the results support the notion that AE sensors could be very valuable for intelligently monitoring the condition of power transformers. These findings could have important implications for the design of automatic power transformer monitoring schemes. In the future, it is hoped that innovations based on these results could be developed and tested, e.g. for deployment in predictive maintenance systems.

References

1. Harbaji, M., El-Hag, A., Shaban, K.: Accurate partial discharge classification from acoustic emission signals. In: 2013 3rd International Conference on Electric Power and Energy Conversion Systems (EPECS), pp. 1–4. IEEE (2013)
2. Brown, M.H., Sedano, R.P.: Electricity transmission: a primer. Natl Conference of State (2004)

3. Kuo, C.C., Shieh, H.L.: Artificial classification system of aging period based on insulation status of transformers. In: 2009 International Conference on Machine Learning and Cybernetics, vol. 6, pp. 3310–3315. IEEE (2009)
4. Sikorski, W., Ziomek, W.: Detection, recognition and location of partial discharge sources using acoustic emission method. Acoustic Emission, pp. 49–74 (2012)
5. Woon, W.L., El-Hag, A., Harbaji, M.: Machine learning techniques for robust classification of partial discharges in oil-paper insulation systems. IET Sci. Measur. Technol. **10**(3), 221–227 (2016)
6. Swedan, A., El-Hag, A., Assaleh, K.: Acoustic detection of partial discharge using signal processing and pattern recognition techniques. Insight-Non-Destr. Test. Cond. Monit. **54**(12), 667–672 (2012)
7. Boczar, T., Borucki, S., Cichon, A., Zmarzly, D.: Application possibilities of artificial neural networks for recognizing partial discharges measured by the acoustic emission method. IEEE Trans. Dielectr. Electr. Insul. **16**(1), 214–223 (2009)
8. Harbaji, M., Shaban, K., El-Hag, A.: Classification of common partial discharge types in oil-paper insulation system using acoustic signals. IEEE Trans. Dielectr. Electr. Insul. **22**(3), 1674–1683 (2015)
9. Hand, D.J., Mannila, H., Smyth, P.: Principles of Data Mining. MIT press (2001)
10. Hastie, T., Tibshirani, R., Friedman, J., Hastie, T., Friedman, J., Tibshirani, R.: The Elements of Statistical Learning, vol. 2. Springer, New York (2009). https://doi.org/10.1007/978-0-387-84858-7

Nonintrusive Load Monitoring Based on Deep Learning

Ke Wang[1,2], Haiwang Zhong[1,2(✉)], Nanpeng Yu[3], and Qing Xia[1,2]

[1] Department of Electrical Engineering, Tsinghua University,
Beijing 100084, China
zhonghw@tsinghua.edu.cn
[2] State Key Laboratory of Control and Simulation of Power Systems
and Generation Equipments, Tsinghua University, Beijing 100084, China
[3] University of California, Riverside, CA 92521, USA

Abstract. This paper presents a novel nonintrusive load monitoring method based on deep learning. Unlike the existing work based on convolutional neural network and recurrent neural network with fully connected layers, this paper develops a deep neural network based on sequence-to-sequence model and attention mechanism to perform nonintrusive load monitoring. The overall framework can be divided into three layers. In the first layer, the input active power time sequence is embedded into a group of high dimensional vectors. In the second layer, the vectors are encoded by a bi-directional LSTM layer, and the N encoded vectors are added up to form a dynamic context vector according to its weights calculated by the attention mechanism. In the third layer, an LSTM-based decoder utilizes the dynamic context vector to calculate the disaggregated power consumption at every time step. The proposed method is trained and tested on REFITPowerData dataset. The results show that compared to the state-of-the-art methods, the proposed method significantly increases the accuracy of the estimation for the disaggregated power value and decreases the misjudge rate by 10% to 20%.

Keywords: Nonintrusive load monitoring · Deep learning
Sequence-to-sequence model · Attention mechanism

1 Introduction

To increase energy efficiency [1] and enhance demand response [2] capabilities of end-use customers, it is crucial to inform household users of real-time electricity consumption of individual appliances in the buildings. Traditional load monitoring methods require a separate sensor for each individual appliance, which results in high implementation cost and low user acceptance. Nonintrusive load monitoring (NILM) is an emerging technology that estimates the electricity usage of individual loads from a single-point measurement of the combined power consumption.

In the literature, most of the research can be grouped into two categories [3]. The first is transient-state-based NILM approach and the second is steady-state-based NILM approach.

© Springer Nature Switzerland AG 2018
W. L. Woon et al. (Eds.): DARE 2018, LNAI 11325, pp. 137–147, 2018.
https://doi.org/10.1007/978-3-030-04303-2_11

Transient-state-based NILM approach utilizes high-frequency electricity features such as voltage and current waveform and its harmonics. Cox [4] conducts frequency spectrum analysis on transient voltage waveform, which reaches high accuracy identifying ON/OFF motion of a single appliance. Tsai [5] applies KNN (K-Nearest Neighbor) and BP-ANN (Back Propagate-Artificial Neural Network) on transient current waveform. Yun [6] carries out similarity matching between standardized template and extracted active and reactive power variation. These methods are based on high frequency sampling and have high requirements for metering devices and data storage systems. Thus, they are not appropriate for household load monitoring.

Steady-state-based NILM approach often utilizes active and reactive power consumption sampled at low frequency. Kolter [7] adopts FHMM (Factorial Hidden Markov Model), and is able to handle situations when multiple appliances are operating simultaneously. Kelly [8] grasps the upsurge of artificial intelligence, and proposed a deep learning model based on CNN (Convolutional Neural Network) and RNN (Recurrent Neural Network). These methods are compatible with the existing smart meter infrastructure. However, as a result of low-frequency sampling, the existing approaches yield high misjudge rate and low accuracy of disaggregated power consumption estimation.

In this paper, we adopt the deep learning framework, and propose a NILM model based on sequence-to-sequence model and attention mechanism. The proposed model is trained and evaluated on the REFITPowerData [9] dataset. The testing results show that the proposed model has great potential in reducing misjudge rate and increasing accuracy of disaggregated power consumption.

The rest of the paper is organized as follows. In Sect. 2, the proposed model and NILM system are introduced. The data source and processing method are presented in Sect. 3. Experiment studies are reported in Sect. 4. Section 5 concludes the paper.

2 Nonintrusive Load Monitoring Model

2.1 Physical Model of Nonintrusive Load Monitoring

Suppose there are M electrical appliances, and the power consumption time series of the ith electrical appliance is:

$$X_i = (x_{i,1}, x_{i,2}, \ldots, x_{i,T}) \, x_{i,t} \in R_+$$

Considering potential random noise in the smart meter measurements, the model can be represented as:

$$y_t = \sum_{i=1}^{M} x_{i,t} + \mu_t \, t = 1, 2, \ldots, T$$

where μ_t is Gaussian noise with zero mean and variance of σ^2.

The aim of nonintrusive load monitoring is to recover the power consumption of each electrical appliance from the aggregated power consumption data. Suppose Y is the aggregated power consumption with T sampling point, and:

$$Y = (y_1, y_2, \ldots, y_T) \, y_t \in R_+$$

2.2 Motivation for Adopting Deep Learning Framework

Deep learning is based on learning data representations, and has made great achievements in computer vision [10], speech recognition and natural language processing [11]. The universal approximation theorem proves that a feed-forward neural network containing finite number of neurons can approximate any continuous functions on compact subsets R^n. This characteristic makes deep learning a great candidate in approximating any real physical model. Hence, it should be suitable for the application of nonintrusive load monitoring.

2.3 Nonintrusive Load Monitoring Model

The overall framework of the proposed nonintrusive load monitoring model is shown in Fig. 1.

The proposed model only utilizes low sampling-rate aggregated active power consumption data, and the data is discretized to integers for simplicity. A deep learning framework with different parameter settings are trained to estimate the load consumption of M different target appliances.

2.3.1 Data Segmentation

Before the disaggregation process starts, data segmentation is conducted to split the long-range input data into pieces of preset length. This preset length is different from each other because different target appliance has different length of run time. For example, microwave's run time is typically less than ten minutes while a washing machine can operate for more than two hours.

2.3.2 Embedding

After data segmentation, an embedding process is used to map the integer value of aggregated power consumption to a high dimensional vector with an embedding matrix E:

$$E = [voc\_size, embedding\_size]$$

For each aggregated power consumption value i, it is mapped to vector $E[i]$, and after the data segmentation process, the $[N_i * 1]$ input sequence is transformed into $[N_i * embedding\_size]$ matrix Z.

2.3.3 Sequence-to-Sequence Model

Sequence-to-sequence model [11] is the encode-decode part in Fig. 1 which converts a sequence to another sequence, and both encoder and decoder are based on LSTM [12]

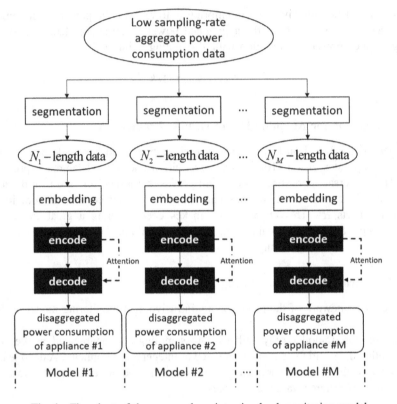

Fig. 1. Flowchart of the proposed nonintrusive load monitoring model

(Long short term memory). The deep learning architecture of sequence-to-sequence model is shown in Fig. 2.

At each time step t, the encoder calculates h_t, the hidden state of time t, from h_{t-1} and Z_t, the embedded input of time t.

$$h_t = f(Z_t, h_{t-1})t = 1, 2, \ldots, N_i$$

where f is the inner computation rule of LSTM. After the encoding process is finished, we can get N_i hidden states, and the last hidden state h_T is assigned to context vector C.

In the decoding process, at each time step t, the decoder calculates s_t (to distinguish between encoder's hidden state h_t), the hidden state of time t, from s_{t-1}, Y_{t-1}, the output of time $t - 1$, and context vector C.

$$s_t = f([Y_{t-1}, C], s_{t-1})t = 1, 2, \ldots, N_i$$

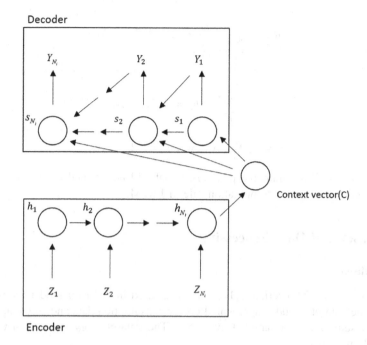

Fig. 2. Schematic of LSTM-based sequence-to-sequence model

Y_t is calculated from s_t:

$$Y_t = g(s_t)t = 1, 2, \ldots, T$$

where g is a projection layer.

2.3.4 Attention Mechanism

One drawback of the sequence-to-sequence model is that the context vector C is fixed during the decoding process. It means that all input information is restricted to a limited length, which might result in data leakage. The attention mechanism can address the problem.

The key component of attention mechanism [13] is to calculate $a_{t,i}$, the weight of the ith hidden state of encoder at time t in decoding, and add the hidden states up according to their own weight to form a dynamic context vector C_t. In this way, the model can pay attention to data that is closely related to decoding at different times. The attention mechanism thus solves the data leakage problem and improves model's ability to extract useful information.

The attention mechanism works as follows:

$$e_{t,j} = V^T * \tanh(W s_{t-1} + U h_j) \quad t, j = 1, 2, \ldots, N_i$$

$$a_{t,j} = \frac{\exp(e_{t,j})}{\sum\limits_{k=1}^{N_i} \exp(e_{t,k})} t,j = 1, 2, \ldots, N_i$$

$$C_t = \sum_{j=1}^{N_i} a_{t,j} h_j t = 1, 2, \ldots, N_i$$

$$s_t = f([Y_{t-1}, C_t], s_{t-1}) t = 1, 2, \ldots, N_i$$

where V, W and U are trainable parameters that will be updated during the training process, and f is the inner computation rule of LSTM.

3 Dataset and Data Processing

3.1 Dataset

We adopt the REFITPowerData [11] dataset released in 2015 for model training and testing. The dataset includes aggregated power consumption data and single appliance power consumption data sampled every 8 s. The dataset is gathered from October, 2013 to May, 2015.

3.2 Selection of Appliances

In the case study, the fridge, TV, microwave, washing machine and dish washer are selected as the five target appliances. This is because they have different working patterns and make up majority of household electricity consumption. For example, fridge simply has on/off mode while washing machine has multiple operation modes with different power consumptions.

3.3 Data Processing

In order to synthetize the training data, we first need to extract load activation [8], which is the working period power consumption data of each target appliance. The parameter setting of the data extraction is shown in Table 1. The extracted load activations are stored properly.

After extracting load activations, the training data are synthetized in three steps.

First, create an all-zero sequence of length N_i, which is shown in Table 1. Then put one load activation of the target appliance into the sequence entirely with 50% probability. The remaining sequence is unchanged with 50% probability.

Second, for appliances except for the target appliance, put one load activation of each into the sequence with 25% probability, and this does not require the load activation to be put into the sequence entirely.

Third, repeat step one and two for K times, and make a training data that includes K pieces of N_i length sequences.

Table 1. Parameters for extracting load activations and sequence size

Load	Max Power (W)	Min Power (W)	Shortest operation time (point)	Longest time below threshold (point)	Sequence length (point)
Fridge	300	20	100	10	400
TV	300	20	100	10	2000
Microwave	NA	20	5	5	50
Washing machine	NA	20	100	10	1200
Dish washer	NA	20	100	10	1000

4   Experimental Results

In order to show the performance of the proposed model, a case with small data size and a case with large data size are both studied.

4.1   Case with Small Data Size

The case with small data size, which is synthetized randomly from load activations that have been extracted, is used to demonstrate the accuracy of disaggregated power consumption directly, and the results are shown in Fig. 3. It can be observed from Fig. 3 that the proposed model can reach a high level of accuracy in estimating the disaggregated power consumption. The model performs exceptionally well for appliances that have predictable work mode and high power consumption such as microwave, washing machine and dish washer. For appliances with low power consumption during working period such as TV and fridge, the predictions of start and end point are slightly less accurate. However, the accuracy of disaggregated power consumption forecast for appliance with lower power consumption is still at a high level.

4.2   Case with Large Data Size

To fully evaluate the performance of the proposed model, four metrics, accuracy, recall, F1-score and mean absolute error are used. These metrics can be calculated as follows.

$$PRE = \frac{TP}{TP + FP}$$

$$REC = \frac{TP}{TP + FN}$$

$$F1 = 2 * \frac{PRE * REC}{PRE + REC}$$

$$MAE = \frac{1}{T_1 - T_0} \sum_{t=T_0}^{T_1} abs(\tilde{y}_t - y_t)$$

PRE is accuracy, REC is recall and F1 is F1 score. TP is the number of true positives, FP is the number of false positives and FN is the number of false negatives. y_t is the actual power consumption of target appliance at time t, and $\tilde{y}_t$ is the disaggregated power consumption of target appliance calculated by the proposed model. MAE is the mean absolute error, which indicates the accuracy of disaggregated power consumption.

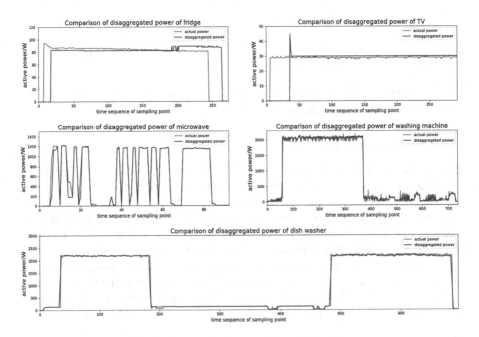

Fig. 3. Test result of case with small data size

To evaluate the performance and capability of generalization of the proposed model, the evaluation of this test case is split into two parts. The first is on houses selected in the training data set, and the second is on houses not included during training. In addition, Kelly's [8] deep learning model is replicated to serve as a benchmark.

4.2.1   Tests on Houses Involved in Training

The test results of houses involved in the training process are shown in Fig. 4.

Tests on houses involved in the training process do not mean the training and testing process is conducted on the same data. In the field of deep learning, the training dataset and testing dataset must be separated because the model may gain some memory of training dataset during the training process. In this work, the training process is on the first one million sampling points and the test process is on the last one

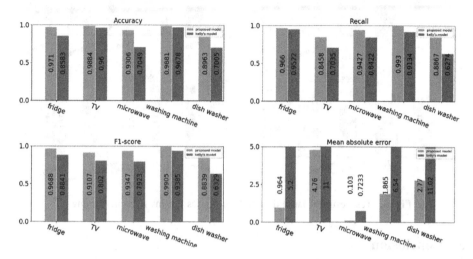

Fig. 4. Test result of large data size on houses involved in training

hundred thousand sampling points.

It can be observed from Fig. 4 that for all appliances, the proposed model can achieve good performance of power disaggregation. The MAE of the proposed method is much smaller than that of the Kelly's [8] model. This demonstrates that the proposed model has strong capability in utilizing data and identifying the working pattern of electrical appliances. Besides, the proposed model also increase accuracy, recall and F1-score by 10 to 20%.

4.2.2 Tests on Houses not Involved During Training

The test results of houses not involved during training are shown in Fig. 5.

Tests on houses not involved during training is of great significance because this is the only way to examine whether the model has learned the right pattern and whether the model has strong capability of generalization. Theoretically, for deep learning models, the more data they are trained with, the better the generalization ability is.

It can be observed from Fig. 5 that the performance of proposed model is slightly worse for houses not involved in training. However, the results are still decent, which demonstrates that the proposed model has strong capability of generalization. Compared to Kelly's [8] model, our proposed model improves accuracy, recall and F1-score by 10 to 20% and reduces the mean absolute error sharply.

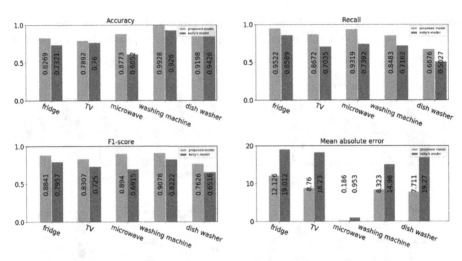

Fig. 5. Test result of large data size on houses not involved in training

5   Conclusion

We develop a deep learning framework based on sequence-to-sequence model and attention mechanism to perform nonintrusive load monitoring. The proposed model introduces the Encoder-Decoder architecture, and uses attention mechanism to extract the most relevant hidden states of encoder to guide the decoding process. These unique features enhance the proposed model's ability to extract and utilize information dramatically. Tests on houses involved in training and houses not involved in training demonstrate that the proposed model can increase accuracy, recall and F1-score by 10 to 20% and reduce mean absolute error dramatically compared to the deep learning models based on CNN and RNN. In the future, we plan to improve the model's capability of generalization and the model's applicability on low power consumption appliances.

Acknowledgement. This work was supported in part by the National Natural Science Foundation of China (51777102), in part by the Beijing Natural Science Foundation (3182017), and in part by the State Grid Corporation of China (5210EF18000G).

References

1. Zhao, J., Dong, C., Wen, F., et al. Data science for energy systems: theory, techniques and prospect. In: Proceedings of the CSEE 2017, vol. 41(04), pp. 1–11 (2017)
2. Zhang, D., Yao, L., Ma, W.: Development strategies of smart grid in China and Abroad. In: Proceedings of the CSEE, 2013, vol. 33(31), 1–15 (2013)
3. Fuqiu, Z., Juan, W.: Situations and suggestions of national DSM work. Power Demand Side Manag. **17**(2), 1–4 (2015)

4. Cox, R., Leeb, S.B., Shaw, S.R., et al.: Transient event detection for nonintrusive load monitoring and demand side management using voltage distortion. In: Applied Power Electronics Conference and Exposition, New York, USA (2006)
5. Tsai, M., Lin, Y.: Modern development of an adaptive non-intrusive appliance load monitoring system in electricity energy conservation. Appl. Energy **96**, 55–73 (2012)
6. Yun, G., Honggeng, Y.: Household load measurement by classification based on Minkowski Distance. Electr. Meas. Instrum. **50**(569), 86–90 (2013)
7. Kolter, J.Z., Jaakkola, T.: Approximate inference in additive factorial HMMs with application to energy disaggregation. In: Artificial Intelligence and Statistics, La Palma, The Republic of Panama (2012)
8. Kelly, J., Knottenbelt, W.: Neural NILM: deep neural networks applied to energy disaggregation. In: Proceedings of the 2nd ACM International Conference on Embedded Systems for Energy-Efficient Built Environments. Seoul, South Korea (2015)
9. Murray, D., Stankovic, L., Stankovic, V.: An electrical load measurements dataset of United Kingdom households from a two-year longitudinal study. Sci. Data **4**, 160122 (2017)
10. He, K., Zhang, X., Ren, S., et al.: Deep residual learning for image recognition. In: Proceedings of the IEEE Conference on Computer Vision and Pattern Recognition, Seattle, USA (2016)
11. Sutskever, I., Vinyals, O., Le, Q.V.: Sequence to sequence learning with neural networks. Adv. Neural Inf. Process. Syst. **27**, 3104–3112 (2014)
12. Hochreiter, S., Schmidhuber, J.: Long short-term memory. Neural Compet. **9**(8), 1735–1780 (1997)
13. Bahdanau, D., Cho, K., Bengio, Y.: Neural machine translation by jointly learning to align and translate. In: Proceedings of the 3rd International Conference on Learning Representations, Leuven, Belgium (2015)

Urban Climate Data Sensing, Warehousing, and Analysis: A Case Study in the City of Abu Dhabi, United Arab Emirates

Prajowal Manandhar[1], Prashanth Reddy Marpu[1], and Zeyar Aung[2(✉)]

[1] Department of Electrical and Computer Engineering,
Khalifa University of Science and Technology, Abu Dhabi, UAE
{prajowal.manandhar,prashanth.marpu}@ku.ac.ae
[2] Department of Computer Science, Khalifa University of Science and Technology,
Abu Dhabi, UAE
zeyar.aung@ku.ac.ae

Abstract. With the ever increasing observations and measurements of geo-sensor networks, satellite imageries, geo-spatial data of location based services (LBS) and location-based social networks has become a serious challenge for data management and analysis systems. In urban micro-climate, we need to deal with various types of data such as: environmental data measurements, Wi-Fi data and so on. The format and the nature of data coming from different sensors such as temperature, humidity, thermal cameras, wind sensors, and others within an urban area varies. Therefore, there is a need for a unified platform to store these data efficiently using new technologies for which, we have come up with implementation of OLAP cubes. Furthermore, additional analytics for assessing urban thermal comfort can also be derived based on behavioural patterns of people. Therefore, outdoor Wi-Fi usage statistics is used as a proxy for the amount of time people spend outdoors, to correlate outdoor thermal conditions to perceived thermal comfort. Some interesting obervations are made in our study.

Keywords: OLAP cubes · Environmental data · Wi-Fi data
Thermal comfort

1 Introduction

We live in a time where a huge amount of sensor based data is being generated and the volume of data is growing with time. Thus, these ever increasing observations and measurements of geo-sensor networks, satellite imageries, and geo-spatial data of LBS has become a serious challenge for data management and analysis systems.

Creating a data warehouse framework allows managing big data, processing large amounts of time series data (big data) and perform analysis by querying

© Springer Nature Switzerland AG 2018
W. L. Woon et al. (Eds.): DARE 2018, LNAI 11325, pp. 148–165, 2018.
https://doi.org/10.1007/978-3-030-04303-2_12

in an optimal way to deal with the data generated by numerous sensors in a micro-climate environment. With growing data, there is a need to store and query data in real time. When the format and the structure of data varies while it comes from different sensors, a need for a unified platform arises to store these data efficiently using latest technologies. Thus, having a single platform for managing the entire urban landscape of an area will have several benefits enabling significant cost savings, causing fewer problems in development phase and providing consistency and manageability in processing the data during the data retrieval phase.

Thus, our work aims to deal with gathering, storage, retrieval, and organisation of data, information and knowledge along with the tools to aid visualization of the stored as well as processed data. Hence, we present a generic OLAP architecture to deal with our data generated from various weather sensor nodes installed in Abu Dhabi downtown area, two of them are shown in Fig. 1. The OLAP cube was also built to store data obtained from the outdoor Wi-Fi routers to relate users' count with respect to environmental variables. Using users' encrypted Wi-Fi data, we analyzed the behavior of people in response to the variations in urban micro-climate. Behavioral modeling is about explaining social phenomena through understanding human behavior in particular systems. Thus, we analyze Wi-Fi data corresponding to different people's behavior (e.g. outdoor activities) and try to model the underlying phenomenon to understand the different behaviors observed under different conditions. Here, we extracted the behavioral traces of human activity that gave rise to social phenomena, producing models of behavior, and integrating behavioral models within systems for predicting macro-scale behaviors.

The main reason for the need of creating this unified platform is that the existing GIS system does not comply with the nature of the data we are dealing with. Hence, our main contribution lies in dealing with the data that possess the following characteristics:

1. Multi-type/multi-structural nodes (each sensor node has a different combination of variables)
2. Varying sampling rate
3. Use of common and uncommon variables

The various environmental data that are collected for data management are temperatures at different heights (4 m, 5.5 m, 6 m, 7.5 m), wind speed, wind velocity, humidity, global horizontal irradiation, ambient/object Temperature, infrared temperatures (mlx90612_1, mlx90621_2, mlx90615_1, mlx90615_2), etc.

Our other contribution lies in the usage of outdoor Wi-Fi statistics to detect the presence of individuals in outdoor spaces rather than using time consuming longitudinal studies and other costly observational approaches [1–4]. Though the positive side of longitudinal studies lies in providing some insight in the 'thermal history' of participants as well as within subject trends. One such example is, a warm day in spring tends to lure more people outside than the similar weather condition in early fall. But the drawback remains in interviewing the same people repeatedly, and then the test subjects may over time develop a bias towards

Fig. 1. Weather sensors installed in Abu Dhabi downtown area

Fig. 2. Outdoor Wi-Fi routers in Masdar City

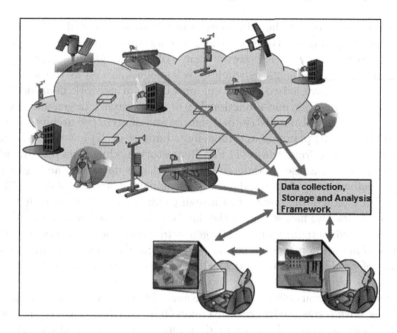

Fig. 3. Overview of data warehousing and analysis framework

the research findings and also, the number of participants is limited for practical reasons. Observational studies using presence counting, such as Gehl [5] studied the influence of micro-climate on outdoor activities by counting people sitting on sunny and shady benches. It showed that sunny or shady conditions significantly impact the desire of people to either stay or leave. But from urban planning perspective, it also matters whether an observed individual is a visitor, or randomly passing independent of prevailing weather, or may be a local resident making the preferable choice to have lunch outside [4]. However, observational approaches could lead towards privacy issues in places like Abu Dhabi, UAE. In addition to this, there could be cost factor associated with the positioning of the cameras as well as synchronization between the cameras to track the people wandering outside. Thus, with the usage of Wi-Fi data, it also allows to perform such studies/analysis in a larger scale. Hence, we have opted to use outdoor Wi-Fi device count as a proxy of number of people being outside (Figs. 2 and 3).

2   Related Works

In this section, we review the existing research in spatial data warehousing and OLAP technology for managing sensor generated data. GIS data is a form of geospatial data. The word geospatial refers the data that has a geographic component to it which means that the records in a dataset have location information tied to them such as geographic data in the form of coordinates, address, city, or ZIP code. Other geospatial data can originate from GPS data, satellite imagery, and geotagging. Most of the GIS tools available today are still utilizing traditional relational formats. However, the new ones have started adopting OLAP and other NoSQL databases, however many still retain SQL-like queries [6].

The recent developments in wireless technologies as well as the widespread usage of sensors have led to the recent prevalence of network monitoring and simulation systems. The main functionality of such systems, in general, is to collect and archive data from distributed sensors and analyze the archived data for applications such as planning and modeling. Jiang et al. [7] has used four module for data storage framework: (1) File Repository module is used for handling small files, (2) Database module combines multiple databases and uses both NoSQL database and relational database for managing structured data, (3) Service module extracts the metadata through configuration, then maps it to both the data entities and files stored in the databases as well as file repository based on the extracted metadata, and finally generating corresponding service; and (4) Resource configuration module supports static and dynamic data management in terms of predefined meta-model, which allows to carry out data disposing mechanism such as load balancing and isolated preferences.

Recently, numerous efforts have been made in the area of data storage and processing, with respect to Internet of Things (IoT). An adopting stream processing techniques is proposed for sensor data integration, a method that supports flexible onboard processing of large volume of sensor data [8]. Li et al. [9] provides a method for wireless sensors to reduce the redundant data collection. Li et al. [10] creates a fast and robust method using posterior-based approximate joint compatibility test to implement data association. Wan et al. [11] proposed five layer system architecture to integrate wireless sensor network (WSN) and RFID technologies. In order to support multi-users to perform data updating or reading, these databases usually sacrifice some features such as database-wide transaction and consistency to achieve higher availability and scalability. In data storage, many traditional data storage platforms are based on relational database. RDMS were used in order to manage and to some extent, analyze the geospatial data in the past. As complements to relational databases, those tools that can efficiently process massive data in distributed environment, that is why modern architecture like OLAP and NoSQL database are getting increasing attention. Though NoSQL databases provide a number of features that relational databases cannot provide, such as horizontal scalability, memory and distributed index, dynamically modifying data schema, etc [12].

In the past, OLAP technology has been used to manage data of weather balloon and flight telemetry system [13]. The framework uses spatial OLAP queries to analyze the trajectory of hurricane with respect to wind speed and direction. Spatial OLAP provides spatio-temporal exploration of historical voluminous data with geovisualization and also, interactivity among the sensors [14]. It enables storage, manipulation and retrieval of analytical queries by means of multi-dimensional expression (MDX). OLAP provides the time based analysis with visual interpretations and also, supports most of the ETL operations.

Moreover, it provides the user with a flexible interactive design to explore the multidimensional spatio-temporal data.

In accordance to the above limitations, a database is required which can handle these different sets of data in urban micro-climate. On the other hand, NoSQL database lacks atomicity, consistency, isolation, durability (ACID) properties and support for some complicated queries. Thus, in this work, we have implemented the OLAP cube as 'Unified Platform' to store and retrieve both environmental data as well as Wi-Fi data.

3   Data Collection and Warehousing

3.1   Data Framework

A typical OLAP architecture comprises of a DataBase Management System (DBMS) to store data (such as MySQL), OLAP server (such as Mondrian Pentaho OLAP Server) that implements the OLAP operation and a browser which combines and synchronizes tabular, graphical, and interactive maps to visualize and trigger desired queries.

Since a traditional DBMS is not able to provide analytical capabilities to analyze and visualize multidimensional data, spatial OLAP is implemented to fully exploit the powerful concepts brought by the multidimensional database structure. This is supported by adding spatial extensions that provide highly interactive map visualization and data exploration. The OLAP approach supports the iterative nature of the analytical process as it allows the user to explore and navigate across different themes (dimensions) at different levels of detail. In addition, it allows rapid visualization of the data at different levels or intersection of dimensions.

The data framework in this study has two components:

1. Data Storage: The backbone of OLAP cube consists of relational database (such as Mysql) which allows efficient processing if maintained at either 2 or 3 normalized form. OLAP can be implemented using either star or snowflake schema. Star schema provides faster performance as the fact table is directly connected with the other dimension tables resulting lesser number of joins in query processing. In the snowflake schema, the dimension tables are also connected to each other, hence, resulting in the need of larger number of joins for querying the data. Star Schema is simpler than Snowflake schema in terms of implementation. Hence, star schema is implemented. The dimension table refers to the table consisting of variables under which the data needs to be explored. For example, the data exploration either with nodes' location-wise or with time-wise. The fact table consists of all the measurable quantities and is linked with the dimension table with a unique 'identifier' field. All the cube information is stored as an XML file. Its graphical representation can be seen in the Fig. 4.

2. Cube Deployment and Data Retrieval: Once the cube is designed, we use the designed XML file and the MySQL connection to deploy our implementation. When the cube is deployed, one can explore the data by means of MDX. Alternatively, one can browse the cube using a browser based Pentaho Analytical tool as shown in Figs. 5, 6 and 7. The main benefits of Cube browsing is that it allows the user to slice and dice the data with a flexibility in data representation by allowing interchange of the dimensional data in either row or column format. While in case of relational reporting tools, it only provides rigid reporting format. It also allows additional drill down and drill (roll) up operations. For the case of reporting, the Pentaho Analytical tool provides easy exportation of desired data in both Excel and PDF formats. The efficiency of the Cube browsing can be increased by means of designing and implementing query based aggregation, or by performing cache warming.

Two cubes has been deployed; one for the case of reporting data and other for the maintenance of the weather sensor nodes that shows the sensor specific information regarding power usage and remaining storage space. The advantage of browsing OLAP cubes is that it allows slicing and dicing the data easily and also enables the user to perform filtering across the multiple variables. The database can handle a variety of queries including a combination of conditions from multiple nodes. For example, a query can be easily performed using filters to provide the output when wind speed at particular Node 1 and Node 13 is higher than 3 m/s. Another advantage of using OLAP cube is that the report does not remain static anymore as with the case of relational report. The nodes of the dimension can be adjusted across the rows and columns as per the users' need.

3.2 Results, Analysis and Discussion

Dataset: We use two types of data for this study. First we use environmental variables such as temperature, wind speed and relative humidity which are obtained with reference from Beam Down, 5 m-Wind Mast weather stations in Masdar city. And to count the number of people, we use encrypted Wi-Fi data obtained from Masdar Institute ICT department. A MacID is a fixed, unique hardware address for a given device such as a smart-phone, tablet or laptop, provided by the manufacturer that supports the device's communication over a network. When in-use, Wi-Fi enabled devices especially smart-phone when in transit, regularly broadcast their MacID in search for better network connections. Given the ubiquitous use of Wi-Fi devices in contemporary urban settings, the number of MacIDs collected at a given place and time strongly correlates with the number of people nearby [15–17]. The encrypted MacID data allowed us to distinguish between Wi-Fi signals from different user groups.

The study area '2A' and '4A' Fig. 10 was chosen as the study site since it is close to sitting areas with shaded and unshaded tables and benches, adjacent to restaurants and cafeterias. Members of Masdar community as well as employees of offices residing in the Masdar City frequently dwell upon these areas especially

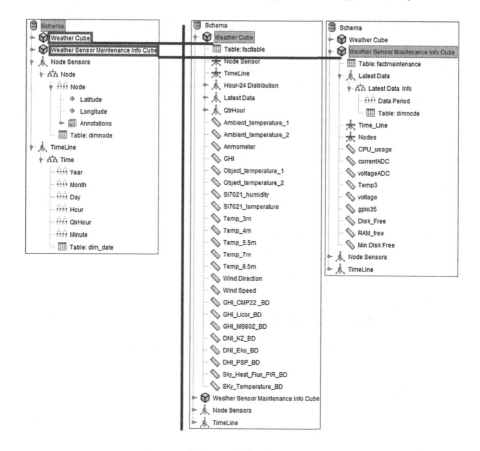

Fig. 4. Cube design

during the lunch hours. We only interpret an outside presence of more than 5 min during lunchtime to constitute an active 'choice' to be outside, postulating that urban workers have an intrinsic bias towards sitting outside when environmental conditions are favorable. Also with reference to Fig. 9, we see that the two top highest counts of people occur nearby the study points which justifies the selection of the study site.

We collected data from 12th May 2016 till 19th April 2017. There was a problem in Wi-Fi data collection from 1st October till 15th November 2016 due to which we had to omit this period making it a total of 229 days excluding weekends and other holidays. The total number of Wi-Fi probes collected during this time period was about 296 thousand out of which 240 thousand were collected during workdays. Using the daily profile of workday probes we further attributed 26 thousand probes as 'lurking' Wi-Fi device. The remaining probes were generated by 3,800 visitors (34 thousand probes) and 1100 regulars (180 thousand probes). We defined a 'regular' as a Wi-Fi user that was present during

24-Hour	A2 Temp_5.5m	E3_3 Temp_5.5m
0	33.413	31.005
1	33.17	30.742
2	32.936	30.546
3	32.687	30.328
4	32.492	30.105
5	32.481	29.984
6	33.123	30.52
7	34.066	31.479
8	35.268	32.735
9	36.736	34.016
10	38.373	34.944
11	39.023	35.409
12	39.061	35.3
13	38.772	34.824
14	38.173	34.395
15	37.324	33.622
16	36.216	32.829
17	35.201	32.337
18	34.6	32.023
19	34.383	31.908
20	34.274	31.79
21	34.098	31.644
22	33.918	31.422

Fig. 5. Cube browsing in tabular representation

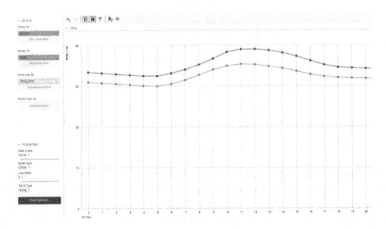

Fig. 6. Cube browsing in graphical representation

at least 10% of all workdays (23 days) throughout the study period. The Wi-Fi probes distribution is shown in Sankey diagram in Fig. 11.

In order to show that Wi-Fi data actually represent the number of people residing in that particular area, analysis of the head count Vs the Wi-Fi device count is performed, by observing the people outside Spinneys canteen area who sat down for more than 5 min. The observation was carried out for more than 1 h 30 min in two different occasions and the two variables showed good correlation as shown in Fig. 14. The counting of the people was carried out both manually

Fig. 7. Cube browsing with interactive map representation

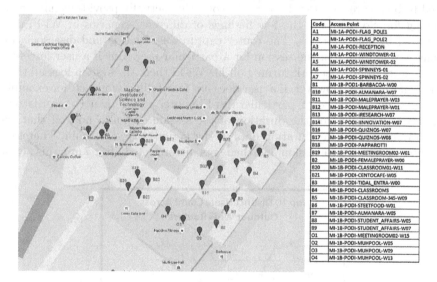

Code	Access Point
A1	MI-1A-PODI-FLAG_POLE1
A2	MI-1A-PODI-FLAG_POLE2
A3	MI-1A-PODI-RECEPTION
A4	MI-1A-PODI-WINDTOWER-01
A5	MI-1A-PODI-WINDTOWER-02
A6	MI-1A-PODI-SPINNEYS-01
A7	MI-1A-PODI-SPINNEYS-02
B1	MI-1B-POD1-BARBACOA-W00
B10	MI-1B-PODI-ALMANARA-W07
B11	MI-1B-PODI-MALEPRAYER-W03
B12	MI-1B-PODI-MALEPRAYER-W01
B13	MI-1B-PODI-IRESEARCH-W07
B14	MI-1B-PODI-IINNOVATION-W07
B16	MI-1B-PODI-QUIZNOS-W07
B17	MI-1B-PODI-QUIZNOS-W08
B18	MI-1B-PODI-PAPPAROTTI
B19	MI-1B-PODI-MEETINGROOM02-W01
B2	MI-1B-PODI-FEMALEPRAYER-W00
B20	MI-1B-PODI-CLASSROOM01-W11
B21	MI-1B-PODI-CENTOCAFE-W05
B3	MI-1B-PODI-TIDAL_ENTRA-W00
B4	MI-1B-PODI-CLASSROOMS
B5	MI-1B-PODI-CLASSROOM-345-W09
B6	MI-1B-PODI-STEETFOOD-W01
B7	MI-1B-PODI-ALMANARA-W05
B8	MI-1B-PODI-STUDENT_AFFAIRS-W05
B9	MI-1B-PODI-STUDENT_AFFAIRS-W07
O1	MI-1B-PODI-MEETINGROOM02-W15
O2	MI-1B-PODI-MUHPOOL-W05
O3	MI-1B-PODI-MUHPOOL-W09
O4	MI-1B-PODI-MUHPOOL-W13

Fig. 8. Location of wireless access point in Masdar Institute

as well as with the Raspberry Pi setup (as shown in Fig. 12). The setup consisted of fish eye lens camera which captured the images every 10 s of the study site (as shown in Fig. 13). And we kept a track of only those people who sat down for more than 5 min.

Figure 15 shows the median occurrence of Wi-Fi probes on workdays throughout the "A (or 1A)" side of Masdar Institute. The "A" site lies in the way towards the Masdar Institute and other residing offices along with the main restaurants and thus shows predictable arrival and departure patterns which rise up from the morning period (around 8 am) until the office time in the evenings (around 6pm). If we consider two main routers in "A" site as shown in Fig. 8 which are; '2A' which is near to Spinneys Canteen and '4A' router which is near to Osha and Sumo restaurants, we see that the number during the lunch hour increases sharply at around 12pm to 2pm and has one more spike at around 4pm which could indicate the tea breaks. The same pattern can be seen in '4A' router side. The regular users on both the routers tend to show similar pattern, the only difference being the number of users. This can be attributed towards the price of Spinneys canteen which is comparatively lower as well as open areas available for sitting infront of '2A' routers. Whereas the price in Sumo and Osha restaurants seemed to be in bit higher side so, the number of users seems to be comparatively lesser on that side. Intuitively, we could even imply that majority of the students try to go towards affordable Canteen while the majority of the employee prefers high priced restaurants.

Observation of Captured Wi-Fi Data: Before we use the Wi-Fi data, we need to be assure that the data we are collecting actually provides what we are looking for. So, here are the few findings of different scenarios of Wi-Fi data captured by Cisco routers. The router captures association and disassociation along with duration of the session for a particular user. But if a user forcefully disconnects from a network turning off his Wi-Fi then, the system logs all these three information. However, if a user does not forcefully disconnect, for example, one goes out of range of particular Wi-Fi router radar then it records only disassociation time not the duration of the session. We also noticed that the Cisco routers store the duration data in the periods of every 5 min. Therefore, if a user connects to a session and then gets disconnected after 7 min, his actual

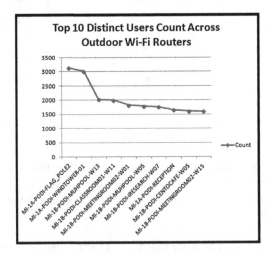

Fig. 9. Top 10 distinct user count across different routers

Fig. 10. Masdar City study site

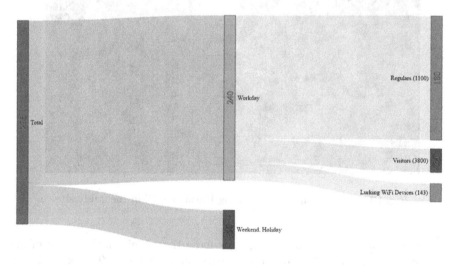

Fig. 11. Sankey diagram showing Wi-Fi probes distribution for routers in "A" site

7 min session is reported in the system as 10 min session. This is also the reason why we neglected the cases of the users who are connected less than 5 min.

We also experimented regarding the capturing of the Wi-Fi session when a user travels from the range of one particular router to the range of another router. For example, if a person gets connected to Wi-Fi from the range of router 'A' and travels to the place which is in the range of router 'C' with another router 'B' in his transit. So, if the total duration of the session is less than 5 min then only one router's information is seen where one spent majority of time of

Fig. 12. Raspberry Pi setup to count the number of people

Fig. 13. Fish-eye camera view using Raspberry Pi in the study site

that session. However, if that user spends more than 5 min in the range of one particular router, then that router's information is also seen. Therefore, in our previous example, if a user travels from a place in a range of router 'A' to router 'C' via router 'B' with the session exceeding 5 min in only two of those routers. Then, the third router's information is not visible in the system. However, the time of the third router is added to one of the other two routers where majority of the time is spent. We also noticed that majority of the sessions were logged into the system with a rounding values of about 5 min which is assumed to be because of the 'session timer or the idle timer' which is set to expire at every 300 s (default value). Thus, if a user leaves Wi-Fi connection without properly disconnecting it from a particular place, his actual 7 min session duration could potentially be seen as 10 min. Moreover, in order to ensure that the Wi-Fi data captured by the outside routers in Masdar Institute were not recording data

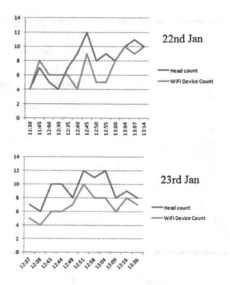

Fig. 14. Head count vs the Wi-Fi device count

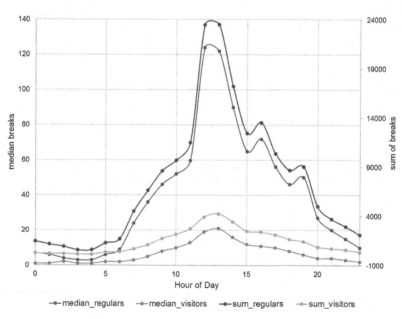

Fig. 15. Median number of Wi-Fi probes on workdays for router in "A" site

from indoor users, we carried out experimentation with the help of IT person from Masdar Institute, which showed that the Wi-Fi devices like smart phones are most likely to get connected to the Wi-Fi router with the highest signal in that particular area. The Wi-Fi signal strength of outdoor routers in an indoor

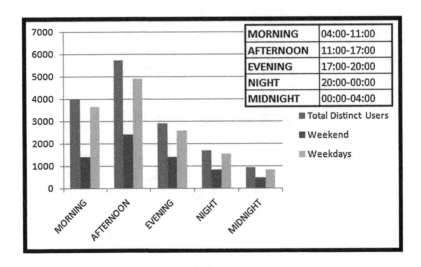

Fig. 16. No. of distinct users at different times of use

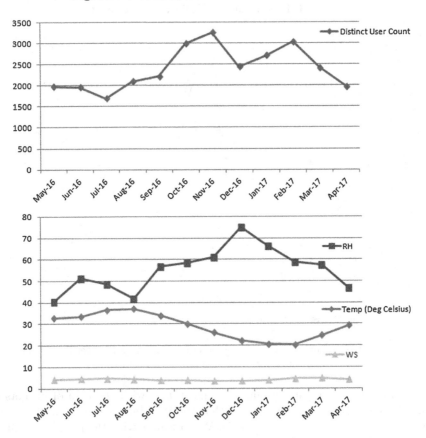

Fig. 17. Distinct users vs average environmental variables

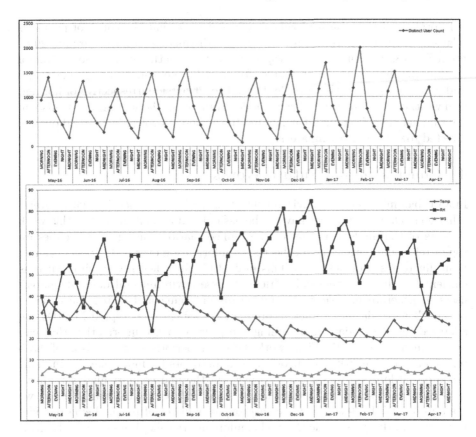

Fig. 18. Distinct users vs average environmental variables in different time of use

environment (mainly besides the closed door) is observed to be very small i.e. received signal strength indicator (RSSI) is found to be less than '−80'. Since Masdar Institute has a routers all over the indoor areas so, the outdoor routers recording the indoor session is very unlikely even if they do, it is advisable to filter the sessions with RSSI greater than '−80' for future use.

With reference to Fig. 16, we observe that the based upon time of use, in overall routers, the distinct count of people using Wi-Fi outdoors appears to be higher in afternoon time, followed by morning, evening, night and mid-night. With reference to Fig. 17, we can say that the average distinct number of people in the summer is about 2000. In the month of July, we see the sudden dip in the count, which could be linked with people taking holidays break. And when the temperature started coming down to the favorable condition (below 30 °C) from the month of October, the number of peoples' count started increasing rapidly till the month of December. In December, we see the dip in the count, which could be linked with the events of semester getting ended and people started going back home for holidays, Christmas and New Year period. Also, when the temperature is favorable till the month of February, we observe that the count of people is relatively high. Then, from April onwards the distinct count hovers around 2000. From Fig. 18, we can see the irrespective of the month, the afternoon time of use has the highest count of people, followed by morning, evening, night and mid-night. This shows that people do tend to sit outside irrespective of hot and cold weather, as there are sufficient shades available in the seating areas. Also, in the summer time, it could be attributed to the desert-coolers being used outside directed towards the sitting areas.

4   Conclusion and Future Works

In an urban micro-climate environment along with remote sensing data, several other local climate variables have to be assimilated as well. A part of this work consists of the implementation of the unified platform using OLAP cubes on top of relational database, to collect such information over wide areas obtained from different sensor networks. This unified platform feature activities such as efficient handling, storing and retrieving these environmental data. It enables users to generate flexible report by means of slicing and dicing the cube as per user's need via its interactive browser. By using an encrypted MacIDs, we analyzed the Wi-Fi data maintaining the privacy of the information that can be obtained from the Wi-Fi network data. We performed comparison of environmental variables with the people sitting outdoors for more than 5 min. This usage of outdoor Wi-Fi statistics to perform thermal comfort analysis, allows to perform such study/analysis in larger scale with a low cost head. Hence, we use outdoor Wi-Fi device count as a proxy of number of people being outside. For future work, the specific thermal comfort metrics like Physiological Equivalent Temperature (PET), Perceived Temperature (PT) and Universal Thermal Climate Index (UTCI) can be studied by performing thermal modeling of the studied area to support our current results of people enduring outdoor thermal weather.

References

1. Johansson, E., Thorsson, S., Emmanuel, R., Krüger, E.: Instruments and methods in outdoor thermal comfort studies - the need for standardization. Urban Clim. **10**, 346–366 (2014)
2. Chen, L., Ng, E.: Outdoor thermal comfort and outdoor activities: a review of research in the past decade. Cities **29**(2), 118–125 (2012)
3. Lin, T.-P.: Thermal perception, adaptation and attendance in a public square in hot and humid regions. Build. Environ. **44**(10), 2017–2026 (2009)
4. Huang, J., Zhou, C., Zhuo, Y., Xu, L., Jiang, Y.: Outdoor thermal environments and activities in open space: an experiment study in humid subtropical climates. Build. Environ. **103**, 238–249 (2016)
5. Gehl, J.: Life Between Buildings: Using Public Space. Island Press, London (2011)
6. Zhang, X., Song, W., Liu, L.: An implementation approach to store GIS spatial data on NoSQL database. In: 22nd International Conference on Geoinformatics (GeoInformatics), pp. 1–5 (2014)

7. Jiang, L., Da Xu, L., Cai, H., Jiang, Z., Bu, F., Xu, B.: An IoT-oriented data storage framework in cloud computing platform. IEEE Trans. Ind. Inform. **10**(2), 1443–1451 (2014)

8. Schweppe, H., Zimmermann, A., Grill, D.: Flexible on-board stream processing for automotive sensor data. IEEE Trans. Ind. Inform. **6**(1), 81–92 (2010)

9. Li, S., Da Xu, L., Wang, X.: Compressed sensing signal and data acquisition in wireless sensor networks and internet of things. IEEE Trans. Ind. Inform. **9**(4), 2177–2186 (2013)

10. Li, Y., Li, S., Song, Q., Liu, H., Meng, M.Q.-H.: Fast and robust data association using posterior based approximate joint compatibility test. IEEE Trans. Ind. Inform. **10**(1), 331–339 (2014)

11. Wang, L., Da Xu, L., Bi, Z., Xu, Y.: Data cleaning for RFID and WSN integration. IEEE Trans. Ind. Inform. **10**(1), 408–418 (2014)

12. Cattell, R.: Scalable SQL and NoSQL data stores. ACM SIGMOD Rec. **39**(4), 12–27 (2011)

13. Viswanathan, G., Schneider, M.: On the requirements for user-centric spatial data warehousing and SOLAP. In: Xu, J., Yu, G., Zhou, S., Unland, R. (eds.) DASFAA 2011. LNCS, vol. 6637, pp. 144–155. Springer, Heidelberg (2011). https://doi.org/10.1007/978-3-642-20244-5_14

14. Rivest, S., Bédard, Y., Proulx, M.-J., Nadeau, M., Hubert, F., Pastor, J.: Solap technology: merging business intelligence with geospatial technology for interactive spatio-temporal exploration and analysis of data. ISPRS J. Photogramm. Remote. Sens. **60**(1), 17–33 (2005)

15. Jiang, S., Ferreira, J., Gonzalez, M.C.: Activity-based human mobility patterns inferred from mobile phone data: a case study of Singapore. IEEE Trans. Big Data **3**(2), 208–219 (2017)

16. Sevtsuk, A., Huang, S., Calabrese, F., Ratti, C.: Mapping the MIT campus in real time using WiFi. In: Handbook of Research Urban Informatics: The Practice and Promise Real-Time City (2009)

17. Freudiger, J.: How talkative is your mobile device? An experimental study of Wi-Fi probe requests. In: Proceedings of the 8th ACM Conference on Security & Privacy in Wireless and Mobile Networks, p. 8 (2015)

Author Index

Printed in the United States
By Bookmasters